Machine Shop Practice, Volume 1

MACHINE SHOP PRACTICE

Volume 1

KARL HANS MOLTRECHT
Technical and Vocational Editor, Industrial Press Inc.

INDUSTRIAL PRESS INC., 200 Madison Avenue
New York, N.Y. 10016

MACHINE SHOP PRACTICE—Volume 1

Copyright © 1971 by Industrial Press Inc., New York, N.Y. Printed in the United States of America. All rights reserved. This book or parts thereof, may not be reproduced in any form without permission of the publishers.

Library of Congress Catalog Card Number: 77–115167

ISBN 0-8311-1069-4

Contents

List of Tables	vii
Preface	ix
1. Basic Metal Cutting Operations	1
2. Basic Measuring Instruments	14
3. Layout Work	44
4. Drilling Machine, Twist Drills and Auxiliary Tooling	61
5. Drilling Machine Operation	90
6. Engine Lathe Construction	108
7. Single-point Cutting Tools and Their Performance	127
8. Cylindrical Turning	168
9. Chucking Work	200
10. Taper and Angle Turning	220
11. Faceplate Work	236
12. Screw Threads and Screw-Thread Measuring	248
13. Cutting Screw Threads on a Lathe	275
14. Turret Lathes, Production Lathes, and Vertical Lathes	304
15. Precision Hole Location—The Jig Borer	323
16. Metal-Cutting Saws	364
Appendix	389
Index	393

List of Tables

2-1.	Accuracy Standards for Precision Gage Blocks	35
2-2.	Sizes for an 83-piece Gage Block Set	37
4-1.	Suggested Lip Relief Angle	69
4-2.	Recommended Drill-point Angles	70
4-3.	Thrust in Pounds for Drilling AISI 1112 Steel	76
4-4.	Oversize Amount Normally Cut by a Drill Under Normal Shop Conditions	77
4-5.	Approximate Conversion Factors for Thrust to Drill Other Materials	85
4-6.	Horsepower at the Cutter Required to Drill AISI 1112 Steel at 100 rpm	86
4-7.	Approximate Conversion Factors, F, for Horsepower at the Cutter Required to Drill Other Materials	86
7-1.	Recommended Cutting Speeds for Turning Ferrous and Nonferrous Metals	129
7-2.	Cutting Speed Correction Factors for Feed and Depth of Cut When Turning	133
7-3.	Recommended Cutting Speeds for Turning Tool Steels	134
7-4.	Chemical Composition of Frequently Used High-Speed Steels	137
7-5.	Recommended Rake Angles	142
8-1.	Unit Horsepower for Turning Ferrous Metals and Alloys	194
8-2.	Unit Horsepower for Turning Nonferrous Metals and Alloys	195
8-3.	Feed Correction Factor, C, for Unit Horsepower	195
12-1.	Coarse-Thread Series, UNC and NC—Basic Dimensions	252
12-2.	Fine-Thread Series UNF and NF—Basic Dimensions	253
16-1.	Recommended Saw-blade Pitch, Cutting Speed and Feed for Hacksaw Operation	366
16-2.	Recommended Cutting Speeds for Band Saw Operation	372

Preface

A very large variety of machine tools is used in modern machine shop practice. Many of these machine tools contain sophisticated devices which are derived from the latest discoveries in science and technology. With the ever-increasing knowledge of science and technology, future machine tools promise to be even more complex to satisfy the demand for more productivity, lower-cost productivity, and greater precision. However, for all of their advantages, the more sophisticated machine tools of today have not replaced the need for the basic machine tools. The basic machine tools are still required to make precision parts for tools, dies, machines, instruments, and for all classes of machine tools.

The best starting point in learning the principles of the more complex machine tools is still to learn the fundamentals of operating the basic machine tools. This can be a stepping stone to a career on the management level in the metalworking industry or as a designer of tools, dies, and machines. It can also lead to a very rewarding career as a skilled craftsman, such as a general machinist, or as a tool and die maker.

This book is intended to treat the operation of the basic machine tools and the technology which supports these operations. The skilled craftsman must be able to operate machine tools intelligently. His ability to advance will, to a large measure, depend upon his ability to apply the supporting technology. The beginning reader may wish to bypass some of the more complicated topics and concentrate at first on the principles of machine-tool operation; however, he will have available for future reference much very valuable technical information on shop theory.

The teacher may likewise wish to bypass the more complicated topics in his course of study when using this book. The inclusion of more advanced topics in this book is an added dimension which can be used in the advanced vocational and technical courses. For the beginning courses, the procedures for operating the basic machine tools are treated in depth. The more advanced lessons can be used to show the extent of the technology of machine shop practice. They can also be used selectively as added assignments for the more ambitious student.

The treatment of the operation of the machine tools in this book makes it possible to use it in the shop. Since it is rarely possible to have all of the students operating the same kind of machine tool at one time, the reading assignments can be made on the basis of the machine tool

that the student is learning to operate. This book can also be used in the related course in which machine shop theory is taught. For this course the subject matter is developed in an orderly and progressive manner. The questions at the end of each chapter can be used for review, for homework assignments, or for examinations.

I would like to take this opportunity to acknowledge the debt of gratitude that I owe to the many persons who have made this work possible. First are the teachers who, in the classroom and in the industrial workshop, taught me the basic skills and principles treated in these volumes and who inspired me to continue my career in this field. It gives me a genuine sense of pleasure to include their names in this book:

<div style="margin-left:2em;">

Orlan W. Boston Clarence J. Metzgar*
Lester V. Colwell John H. Moltrecht*
William W. Gilbert Karl Schatz
Herman Hermann G. Harold Silvius
Frank Kiefer* Robert E. Smith*
Richard O. Knight William C. Truckenmiller
Robert E. McKee Carl Wolff*

</div>

deceased

The acknowledgments in the captions of the illustrations give some idea of the help offered by the machine tool, and allied industries in the preparation of this book. The generous help and cooperation of the persons and their employers who furnished this material is most sincerely appreciated. It is with deep regret that it is not possible to list here all of these individuals, and their firms, by name.

I must also express my appreciation to Holbrook Horton of the Industrial Press Inc. His suggestions and corrections contributed very materially to the clarity and accuracy of these volumes.

Lastly, I am much indebted to my wife, Wilma, for her constant help and encouragement. To her I tender my very best thanks for deciphering the original handwritten manuscript and for toiling endless hours at the typewriter in the preparation of this book.

CHAPTER 1

Basic Metal Cutting Operations

This treatise deals with the principles, tools, and processes which are utilized in all types of machine shops or machine building plants. The information is largely fundamental in character, and the various tools and processes described are applicable in making many different classes of machines and other mechanical products. While machine building, in a broad sense, embraces every craft and process from the making of patterns, castings, forgings, stampings, etc., to the exact procedure in the finishing and assembling of parts, this course of instruction applies principally to the work of the machine shop—because of the general demand for such information—and its usefulness in vocational schools and in the industrial training departments of manufacturing plants. The application of different types of machine tools and machining operations is explained. It will be noted that many of the operations described can be performed in the same, or very similar manner, on different machine tools. For example, boring is done on a lathe, a horizontal boring machine, a drilling machine, a milling machine, and a jig borer. An awareness of these similarities is very helpful in mastering the skills and the related knowledge required to operate machine tools.

The Work of the Machine Shop

The work of the machine shop in building machines is to make each part to the required shape and size. The unique feature of all machine shop operations is that the parts are precisely formed to the desired shape and size by removing metal from the workpiece in the form of chips. These metal chips may be very large, or, as in the case of grinding operations, very small. Assembling of the finished parts is, of course, the final job. The material which is to be machined is generally in the form of hot-rolled bars, cold-drawn bars, centerless ground or machined bars, castings, or forgings. Tools, dies, machine parts, and engine parts frequently must have very precise dimensions and smooth surface finishes. These characteristics can sometimes be obtained by other manufacturing methods, but they can usually be achieved more economically, and to a higher degree, by machining. Value is added to the raw materials that are processed to become machine parts in the machine shop, by virtue of the work done to change their shape, provide precise dimensions, and create their final, smooth surface finishes.

Machine Tools

The general term "machine tool," is applied to various classes of power-driven metal cutting machines such as are used in machine shops. Most machine tools change the shape of a material by producing chips; however, the National Machine Tool Builders' Association includes power-driven presses in the category of machine tools even though they do not produce chips. In these two volumes, however, power-driven presses, such as are used in the stamping and forging industry, will not be considered since, generally, they are not used in the average machine shop.

Machine tools can be considered to serve four main purposes:

1. They hold the work or the part to be cut.
2. They hold the cutting tool or tools.
3. They impart to the cutting tool or work—or both—whatever motion is required for cutting or forming the part.
4. They are arranged, ordinarily, for regulating the cutting speed and also the feeding movement between the tool and the work.

In the production of machine parts of various shapes and sizes, the type of machine and cutting tool used will depend upon the nature of the metal-cutting operation, the character of the work, and, possibly, other factors such as the number of parts required and the degree of accuracy to which the part must be made. The development of machine tools has been largely an evolutionary process, as they have been designed to produce parts meeting increasingly stringent mechanical standards. Developments in power transmission, accuracy, and control of the movements and functions of the machine are constantly being incorporated into the design of new machine tools.

Machine tools are built to produce certain shapes such as cylindrical and conical surfaces, holes, plane surfaces, irregular contours, gear teeth, etc. Many machines, however, can produce a variety of surfaces. Thus, machine tools are built as general purpose machines, high production machines, and as special purpose machines. As the name implies, general purpose machine tools are designed to be quickly and easily adapted to a large variety of operations on many different kinds of parts. They are used extensively in jobbing shops, repair shops, and in tool and die shops. Sometimes these machine tools are called job shop machine tools. Production machine tools are designed to perform an operation, or a sequence of operations, in a repetitive manner in order to achieve a rapid output of machined parts at minimum cost. They can be set up to machine a variety of different parts; however, their operation is economical only when the quantity of parts to be machined is relatively large. Special purpose machine tools are designed to perform one operation, or a sequence of operations, repetitively, on a specific part. These machines are usually automatic and are unattended except when it is necessary to change and to adjust the cutting tools. They are used in mass-production

shops such as are found in the automotive industry. Numerically controlled machine tools have their functions programmed on a punched tape. The tape may be made on a tape writer in accordance with a program written to produce the required part or it may be made from a program determined by a computer. Numerically controlled machine tools achieve a rapid and economical output of machine parts when the quantities to be produced are low or medium. They cannot cut metal any faster than can other machine tools; however, they reduce the amount of "cut and try" by the machine operator thereby increasing the output of parts. They are particularly useful where repeat orders for the same part are encountered. The tape containing the information required to produce the part can be stored conveniently and reused when required.

Metal Cutting Tools

The metal cutting tool separates chips from the workpiece in order to cut the part to the desired shape and size. There is a great variety of metal cutting tools each of which is designed to perform a particular job or a group of metal cutting operations in an efficient manner. For example, a twist drill is designed to drill a hole having a particular size, while a turning tool might be used to turn a variety of cylindrical shapes. In order to sever a chip from the workpiece the following conditions must be present:

1. The tool is harder than the metal to be cut.
2. The tool is so shaped that its cutting edge can penetrate the work.
3. The tool is strong and rigid enough to resist the cutting forces.
4. There must be movement (provided by the machine tool) of the tool relative to the material in order to make the cutting action possible.

Modern metal cutting tools are made from tool steels, powdered metals, ceramics, and industrial diamonds. These materials can be made to be very hard and they can retain their hardness at the high temperatures resulting from the metal cutting action. All metal cutting tools wear as the result of stresses and temperatures encountered in separating the chips. The rate-of-wear must be controlled by the application of the correct cutting speed and feed. After wear has progressed to certain limits the cutting edge may be resharpened by grinding. Ultimately, further sharpening is not practical and the tool must be discarded. For this reason metal cutting tools are classified in industry as perishable tools.

There are three basic types of metal cutting tools: single-point tools, multiple-point tools, and abrasives. These names are quite descriptive. A single-point metal cutting tool has a single cutting edge and is used for turning, boring, shaping, and planing. Multiple-point tools have two or more cutting edges such as drills, reamers, and milling cutters. Grinding wheels are an example of abrasive cutting tools. Each grinding wheel has thousands of embedded abrasive particles which are capable of pen-

Principles of Turning

etrating the workpiece and removing a tiny chip. The combined total of the tiny grinding chips can result in a substantial amount of metal being removed from the workpiece.

Principles of Turning

The diagram at the left in Fig. 1-1 illustrates how metal is cut by turning, as in a lathe. The turning tool is held rigidly in some form of holder. The part to be turned rotates in direction R, and the tool feeds along continuously in direction F, parallel to the axis of the part being turned. Consequently, the part is reduced from some diameter D, to some other diameter d, depending upon the requirements in each case. The upper surface of the tool, against which the chip bears as it is being severed, usually is ground to some rake angle a, to make the edge keener and to lessen cutting resistance. (There are exceptions to this rule as explained later in Chapter 7 which covers Single-Point Tool Forms and Tool Grinding.) This top or chip-bearing surface generally slopes away from that part or section of the cutting edge which normally does the cutting; hence, usually there is both backward and side slope or, in tool parlance, a back rake angle and a side rake angle. (The angle, a, of the diagram is

Fig. 1-1. Principles of turning and planing.

intended to represent the combined back and side angles or what is known as the compound angle in some plane x-x.) If this inclination a, is excessive, the tool point will be weakened. On the other hand, if there is no inclination, the tool will still remove metal, but not as effectively as a tool with rake, especially in cutting iron and steel. Below the cutting edge, some relief angle c, is essential so that this end or side surface of the tool will not rub against the work and prevent the cutting edge from entering it freely, particularly when the tool is adjusted inward for taking a deeper cut. This relief is not confined to the end alone but is also required along the side extending as far back as the length of the cutting edge. The factors governing the amount of rake angle a, relief angle c, and the general shape of the cutting edge will be considered later.

Principles of Planing and Shaping

A comparison of the turning-tool diagrams (Fig. 1-1, left) with the planing tool (Fig. 1-1, right) will show that planing a flat surface is similar in principle to turning a cylindrical surface, the chief difference being that a straight-line movement is required for planing and a rotary or circular motion is used for turning. (The planing tool, when planing a horizontal surface, is held in a vertical position. It is shown in a horizontal position to obtain a more direct comparison with the turning tool.) The planing tool is held stationary while cutting and the part to be planed is given a straight-line motion R, during the cutting stroke, followed by an intermittent, lateral tool-feeding movement in direction F, during the return or non-cutting stroke. (If a shaper, which is a small type of planing machine, is used, this order is reversed; the tool moves, while the work remains stationary.) As the tool is moved laterally, prior to each cutting stroke, successive strokes reduce the surface of the work from H to h, the amount depending upon the requirements.

In planing, as in turning, the tool surface against which the chip bears usually slopes backward or has a certain rake angle a, to reduce the cutting resistance. Some relief angle c, in back of the cutting edge, is necessary to permit this edge to cut freely. If either angle a or c is excessive, the point of the tool will be weakened and it may not be strong enough to resist the pressure required for cutting iron or steel. This is true of turning, planing, and other classes of metal cutting tools. The shapes and angles of planing tools are dealt with in Volume 2, in the chapters on Shapers and Planers.

Principles of Milling

The milling cutter represented by the upper left diagram, Fig. 1-2, has equally spaced teeth around its circumference so that as many cuts are taken per cutter revolution as there are teeth, and the cutting action is continuous. The cutter is rotated by the spindle of the milling machine and the part to be milled is given a feeding movement, generally in direction F, or *against* the cutter rotation R. (Sometimes the feeding movement is in the opposite direction, or *with* rotation R.) As the work feeds

6 BASIC METAL CUTTING OPERATIONS Ch. 1

Fig. 1-2. Principles of plain milling and face milling.

past the revolving cutter, the surface H, is reduced to some height h, as may be required.

The milled surface will be flat if the cutter is cylindrical, but, frequently, cutters of other forms are used, as explained later in Chapter 6, Volume 2, the section on Milling Machine Operations. The chip-bearing surfaces of the milling cutter teeth may have some rake angle a, to lessen cutting resistance and they must have relief c, to permit free cutting action. Thus we see that each tooth depends for its cutting action upon a relief angle and usually a rake angle also, just as with the turning and planing tools. (The actual relief angle c, applies to a narrow top edge on each tooth as indicated by the enlarged view, at upper right, of one tooth. Normally, the cutter is sharpened by grinding this narrow edge or "land" as it is called.)

The lower diagram in Fig. 1-2 illustrates the action of a face milling cutter. This general type of cutter is designed especially for milling flat surfaces. To simplify the diagram, a cutter is shown with two cutting blades only. In actual practice, however, there would be a number of blades equally spaced around the circumference of the cutter body to obtain more continuous cutting action. As the cutter rotates in direction R, the work feeds past it as indicated by arrow F, so that each blade, as it sweeps around, reduces the surface from some height H, to h. The feeding movement of the work is not always in a straight line but may be rotary. This, however, does not affect the operating principle. Each blade is held in a cutter body at an angle as indicated by the detail view to the right. This inclined position gives each blade a certain rake angle, a. The lower

Ch. 1 BASIC METAL CUTTING OPERATIONS 7

end of each blade also has a relief angle c. The face milling cutter is another illustration of the fact that metal cutting operations with different types and designs of tools are all based upon the same fundamental principles.

Principles of Boring

The machining of internal cylindrical surfaces, when a single-point tool is used, is called "boring." Figure 1-3 illustrates the process. The part to be bored rotates in direction R, while the boring tool, with its top surface located in approximately the same plane as the center of the hole, feeds in direction F, or parallel to the axis of the hole, for cylindrical boring. As the tool advances, it increases the diameter of the hole, or bore, from d to D. The cutting tool, when applied to iron or steel, normally would

Fig. 1-3. Principles of enlarging or boring a hole.

have some slope or rake angle a on the chip-bearing surface, and a relief angle c, is essential. Note that there is both backward and side slope (or rake), and also relief at both the extreme end and along the leading side, or below all of the cutting edge. A somewhat greater relief angle may be required for boring than for turning—especially if the hole is small—but the cutting action is based upon the same fundamental principle in each case. A boring tool, of the general type shown in the diagram, is merely a small turning tool supported by a bar B, which, in turn, is rigidly supported at its outer end, and is long enough to permit boring a hole as deeply as may be required.

Principles of Drilling

When a hole is drilled by using a twist drill, the cutting point, or end, is ground to provide a certain amount of relief c, in back of each cutting

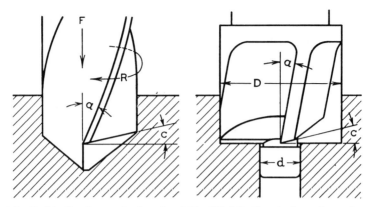

Fig. 1-4. (Left) Cutting end of twist drill. (Right) Counterbore for enlarging a previously drilled hole.

edge (left, Fig. 1-4) and extending along the entire length of the two edges. The twisting, or helical-shaped flute, or groove through which the chips pass also provides rake angle a above each cutting edge. This rake, or backward slope, is away from the cutting edge and is also applied to the chip-bearing surface as in the case of the other tools. It will be understood that the drilling of a hole is done either by feeding a rotating drill in direction F, and into stationary work as when using a drilling machine, or by rotating the work itself instead of the drill as, for example, when using a lathe or turret lathe for drilling.

The diagram, Fig. 1-4, right, illustrates a counterbore. This is a tool for enlarging part of a hole from some diameter d, to D. A pilot, or extension, fits into the smaller hole and insures cutting the enlarged part concentric with it. Again we have on the counterbore, relief angle c, at the lower end of each cutting tooth, and also rake angles a. While the counterbore and the twist drill are in appearance unlike a milling cutter or turning tool, all of these tools conform to the same fundamental operating principles.

Principles of Broaching

A broaching tool (Fig. 1-5) has a series of teeth so arranged that they cut the metal when the broach is given a straight-line movement, as indicated by arrow R. (In some cases the workpiece is moved instead of the broach, but the operating principle is the same.) The broach cuts away the metal since its teeth progressively increase in height, as shown on the exaggerated scale in the lower diagram. Thus, cutting action is distributed over a series of teeth and each tooth takes a light cut.

Each tooth is similar to a planing tool, and the cutting of metal in broaching is based upon the same fundamental principles as in planing, turning, or milling. The rake angle a, as in the case of other metal cutting tools, improves the cutting action. The top, or land, of each tooth has a

Ch. 1 BASIC METAL CUTTING OPERATIONS 9

clearance, or relief angle c. When the broach is initially ground to the size and shape required, a narrow part of this land is often ground without any relief so that precise measurements may be taken on this surface. Later, the relief angle is finish ground, leaving a very small portion of the land behind the cutting edge without a relief angle in order not to destroy the size and shape of the broach. This straight section of the land sometimes helps to stabilize the broach when it is cutting and it provides a slight burnishing or finishing action on the surface of the workpiece. This lack of clearance, or relief, on broach teeth may seem opposed to the principle that relief is an essential requirement; however, the width of the straight portion is usually less than .030 inch. It does not interfere

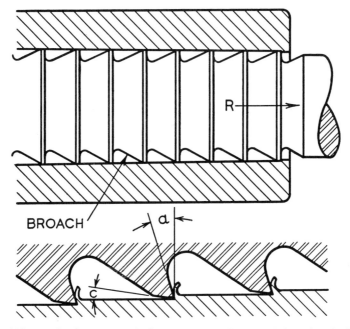

Fig. 1-5. (Above) Cutting action of a broaching tool. (Below) Enlarged scale drawing of broach teeth showing height increases.

with the cutting action because each tooth removes only a small amount of metal and the cutting edges of the broach are never adjusted inward or sunk more deeply into the metal while cutting.

The diagrams show that the cutting action of the broach follows the same principle as the turning tool, planing tool, milling cutter, and other metal cutting tools. Thus, it can be seen that the basic requirements of tool shapes are not complicated. In actual shop practice, however, it is essential to use tools which more or less vary in shape to suit different cutting conditions. The different factors to be taken into account will be referred to later.

10 BASIC METAL CUTTING OPERATIONS Ch. 1

Courtesy of Cincinnati Milacron

Fig. 1-6. Principle of cylindrical grinding, "traverse" method.

Precision Grinding

Grinding wheels are just as much metal cutting tools as are single- or multiple-point cutting tools which are used to alter the shape of a workpiece. Some grinding wheels are used to rough grind castings or steel bar stock in foundries or steel mills. These wheels must be strong enough to withstand rough usage. In the machine shop, however, grinding wheels are usually mounted on the spindles of precision machine tools in order to machine parts to close tolerances and to produce fine surface finishes. Grinding wheels are also used in the machine shop to grind cutting tools on pedestal grinders or on special cutter- and tool-grinding machines. Precision grinding machines are designed to control the path followed by the work and the wheel, with great precision. These machines thus have the capability of producing parts with a high degree of dimensional accuracy.

This capability is further enhanced by the nature of the cutting action of the grinding wheel which produces a multitude of very small chips.

These small chips permit very small amounts of metal to be removed from the surface of the work, thus favoring both dimensional accuracy and surface finish. Another important characteristic of the grinding wheel is that it will readily penetrate hardened metals, such as hardened tool steel, enabling these materials to be machined economically on precision grinding machines. Some heavy-duty surface-grinding machines have been designed to achieve rather high, metal removal rates. This process is called abrasive machining.

Cylindrical grinding machines have been designed to produce ground cylindrical, tapered, and formed surfaces. When traverse grinding (Fig. 1-6), the rotating workpiece is traversed past the grinding wheel which is rotating at a very high speed. At the end of each traverse stroke there is a dwell before the direction of the traverse is reversed. The purpose of the dwell is to give the grinding wheel time to cut at the end of the stroke. Generally, the dwell is also utilized to feed the grinding wheel into the workpiece for the succeeding traverse. This feed is called "infeed." Another method of cylindrical grinding is called "plunge grinding" (Fig. 1-7). This method is limited, however, to surfaces on the workpiece that are no longer than the width of the grinding wheel. The grinding wheel is fed, or plunged, into the rotating workpiece without any traverse motion.

Courtesy of Cincinnati Milacron

Fig. 1-7. "Plunge grinding" method of cylindrical grinding principle.

12 BASIC METAL CUTTING OPERATIONS Ch. 1

Surface grinding is accomplished by traversing a non-rotating workpiece, usually in a horizontal position, past the grinding wheel as shown in Fig. 1-8. The grinding wheel is generally fed a small amount perpendicular (crosswise) to the direction of the traverse movement of the work

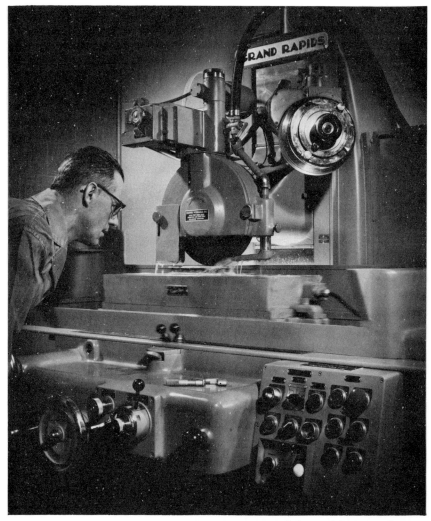

Courtesy of the Gallmeyer & Livingston Company

Fig. 1-8. Principles of surface grinding.

at the end of each traverse stroke. The surface resulting from this action is a plane, or flat surface. Special methods are used to grind profiled surfaces such as threads and gear teeth.

BASIC METAL CUTTING OPERATIONS

QUESTIONS

1-1. Is an awareness of similarities in the operation of different machine tools helpful in mastering the trade?
1-2. What are the advantages of machining, over other metal-working processes?
1-3. What is the main function of the machine shop?
1-4. What are machine tools?
1-5. What is the difference between general purpose machine tools, production machine tools, and special purpose machine tools?
1-6. Name the three basic kinds of metal-cutting tools.
1-7. What is the basic difference between turning and planing?
1-8. Describe two different methods of milling a plane surface.
1-9. What is the difference between boring a hole and drilling a hole?
1-10. Name three basic characteristics of the precision grinding process.
1-11. Describe two methods of performing the cylindrical grinding operation.
1-12. How are flat surfaces ground?

CHAPTER **2**

Basic Measuring Instruments

Measuring tools are used in machine construction, tool and die making, and machine tool operation in order to secure the required sizes and degree of accuracy of the parts. There are many different types and designs of measuring tools, some of which are very sophisticated. This section will deal with the measuring tools normally used by machinists and toolmakers in the everyday performance of their trade. A representative sample of these tools is shown in Fig. 2-1.

All precision measuring tools must be handled with care in order to avoid the slightest damage to them. Often measurements accurate to one-thousandth (.001) inch, or even to one ten-thousandth (.0001) inch, must be made. Careless handling of precision measuring tools will destroy their accuracy, and hence their usefulness, in making exact measurements. For example, a square that is not a true square will result in spoiled work. Precision measuring tools, therefore, must frequently be checked in order to verify their accuracy. With many of these tools the workman must depend upon the sense of feel in order to make accurate measurements. This requires that they be correctly grasped by the hands in order to utilize the sensitive nerves that are located in the tips of the fingers. The proper manipulation of precision tools is shown in many of the illustrations in this chapter.

The Precision Steel Rule

The rule is a basic length-measuring tool. It may be used directly in measuring a length; or, it may be used indirectly as when the diameter of a cylindrical object is measured with outside calipers and this measurement is then transferred from the calipers to the rule. Only precision-made, hardened steel rules are acceptable for use in the machine shop or in the tool and die shop. The lines or graduations on these rules are machine divided on specially designed, and very precise, graduating machines. Precision steel rules may be obtained as spring temper, semi-spring temper, semi-flexible, or full flexible. They range in size from one-quarter inch in length for measuring in grooves, recesses, and keyseats to twelve feet in length for large work. The most frequently used lengths are the six-inch and twelve-inch rules.

Two systems of graduations (Fig. 2-2) are used in the United States on steel rules. They are the decimal inch system and the fractional inch

Ch. 2 BASIC MEASURING INSTRUMENTS 15

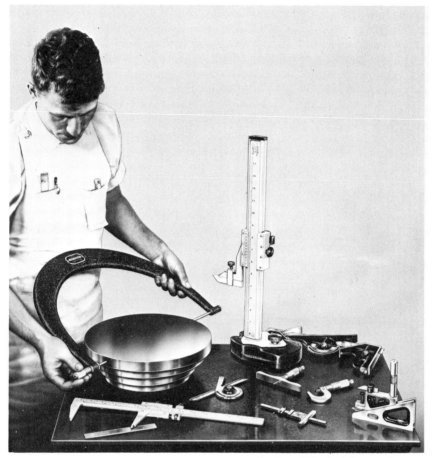

Courtesy of The L. S. Starrett Company

Fig. 2-1. Basic machinists' measuring tools including a precision surface-plate.

Courtesy of The L. S. Starrett Company

Fig. 2-2. (Upper view) Steel rule with decimal inch graduations; (Lower view) Steel rule with fractional inch graduations.

system. In the decimal inch system, the inch is divided into one tenth (.100) inch, one-fiftieth (.020) inch, and one-hundredth (.010) inch. (See upper view, Fig. 2-2.) In the fractional inch system (lower view, Fig. 2-2), the inch is successively divided by two, yielding the following graduations: $\frac{1}{2}$ or .500 inch, $\frac{1}{4}$ or .250 inch, $\frac{1}{8}$ or .125 inch, $\frac{1}{16}$ or .0625 inch, $\frac{1}{32}$ or .03125 inch, and $\frac{1}{64}$ or .015625 inch. One advantage of the fractional inch system is that the smallest graduation that can be clearly distinguished with normal eyesight, and without causing undue eyestrain, is $\frac{1}{64}$ or .0156 inch as compared to $\frac{1}{50}$ or .020 inch for the decimal inch system.

Decimal equivalents are the decimals corresponding to the fractions in the fractional inch system. The decimal equivalent can be calculated from the fraction by dividing the denominator into the numerator. For example, to find the decimal equivalent of $\frac{7}{32}$ inch, divide 7 by 32, thus $7 \div 32 = .21875$ inch. Tables of decimal equivalents are usually available, making these calculations unnecessary. Their use is recommended to prevent possible mistakes that sometimes are made when decimal equivalents are found by calculation. A table of decimal equivalents is given in the Appendix of this text.

Calipers

There are three basic kinds of calipers: 1. Outside calipers, used primarily to measure outside diameters; 2. Inside calipers, used primarily

Courtesy of The L. S. Starrett Company

Fig. 2-3. (Left) Outside calipers; (Center) Inside calipers; (Right) Hermaphrodite caliper.

Fig. 2-4. Correct method of using outside calipers.

to measure hole diameters; and 3. Hermaphrodite calipers, used to measure the distance between a surface and a scribed line or to scribe a line from a surface. Two types of construction, firm joint and spring, are used in calipers. Firm joint calipers utilize the friction between the legs to maintain their setting. A firm joint hermaphrodite caliper is shown in Fig. 2-3. Spring calipers have a curved spring at the upper, or pivot, end which forces the caliper legs against the adjusting screw nut, thereby maintaining the caliper adjustment.

Accurate measurements are obtained with outside calipers by holding them as shown at A, in Fig. 2-4, and passing them over the workpiece with a very light contact pressure that can just be felt by the fingertips. The caliper reading is then transferred to a rule (Fig. 2-4, B). Accurate caliper readings, however, cannot be made when the work is rotating. In measuring a hole, inside calipers are adjusted to the correct diameter of the hole (see A, Fig. 2-5) by holding one caliper leg against the side of the hole with a finger of one hand, while the caliper, held in the other hand, is carefully passed through the hole. Moving the free leg back and forth in the X direction (A, Fig. 2-5), while slowly passing through the hole in the Y direction, will assist in finding the smallest true diameter of the hole through which the caliper legs can pass. Again, a very light feel will determine this setting. The inside caliper measurement is then transferred to a steel rule (B, Fig. 2-5) or to an outside micrometer caliper (D, Fig. 2-5). When done with great care, measurements accurate to

Fig. 2-5. Correct method of using inside calipers.

Courtesy of The L. S. Starrett Company

Fig. 2-6. Telescoping gages used to measure hole diameters.

"tenths" (.0001 inch) can be made, using both inside calipers and outside micrometer calipers. Measurements can also be transferred between inside and outside calipers as shown at C, in Fig. 2-5.

Telescoping-Gage and Small Hole Gages

The telescoping gage, Fig. 2-6A, has two contact plungers which expand outwardly across the diameter of a hole when the knurled nut at the end of the gage stem is released. When both plungers have contacted the hole, Fig. 2-6B, this nut is again tightened and the telescoping gage measurement is transferred to outside micrometer calipers. Small hole gages (Fig. 2-7) have two contact points that are expanded by turning the knurled nut on the end of the stem until the gage can pass through

Courtesy of The L. S. Starrett Company

Fig. 2-7. Small hole gage-setting measured by ouside-micrometer caliper.

the hole with a very light contact "feel." The hole measurement is then transferred to outside micrometer calipers as shown in the illustration.

The Square and the Bevel Protractor

A frequently encountered angle in machine shop work and in toolmaking is the right angle, or 90 degree angle. The square is the basic tool used to measure and test for this angle. Other angles are measured with a protractor, illustrated in Figs. 2-8 and 2-16.

A combination set is shown in Fig. 2-8. Attached to a precision steel rule, from left to right, are a square head, a bevel protractor, and a center head. These heads can be changed to any desired position along the length of the rule or they may be removed completely. The square

Courtesy of The L. S. Starrett Company

Fig. 2-8. Combination set including square head, bevel protractor, and center head.

head and the bevel protractor both contain spirit levels which are often of assistance in making measurements. The square head has a short edge machined at 45 degrees with respect to the right angle. The use of the center head will be described in the next chapter.

For very accurate determinations of the right angle, a master precision square, Fig. 2-9, is used. The beam and the blade of this square are hardened, ground, and lapped to insure straightness, parallelism, and the exact, right-angle relationship. It should be very carefully handled at all times. Squares may be checked by using two toolmakers buttons that are screwed to an angle plate as shown in Fig. 2-10. The lower button is screwed tight, while the other button is tightened also, but less firmly. The square and the angle plate are then set on a precision-surface plate and finally, the blade of the square is placed against the buttons. A thin piece of paper, or paper "feeler," is placed between the blade and each button. The upper button is then lightly tapped with a piece of soft metal, say a 1-inch dia. × 6-inch piece of soft brass, until both paper feelers are tight. The square is then positioned on the other side of the buttons and checked with paper feelers as before. If both paper feelers are tight the square is accurate. In making this test it is very important that both toolmakers buttons have exactly the same diameter. The thin paper

Courtesy of The L. S. Starrett Company

Fig. 2-9. Master precision-steel square.

feelers are often much more sensitive than eyesight when making precise angular measurements.

Extremely accurate right-angle measurements can be made with the cylindrical square shown in Fig. 2-11. It is used on a precision-surface plate. This square is a perfectly true cylinder having one end lapped exactly at right angles to the cylindrical surface. The other end is lapped to a fixed angular relationship to the sides. When the angular end is resting on the surface plate and the work is in contact with the cylindrical surface, the deviation from the true square condition of the part can be measured by rotating the square until the light visible between the part and the square is shut out. The deviation is read by reading up the topmost dotted curve in contact with the work to the number which shows the deviation in steps of .0002 inch.

Fig. 2-10. Method for checking a square.

22 BASIC MEASURING INSTRUMENTS Ch. 2

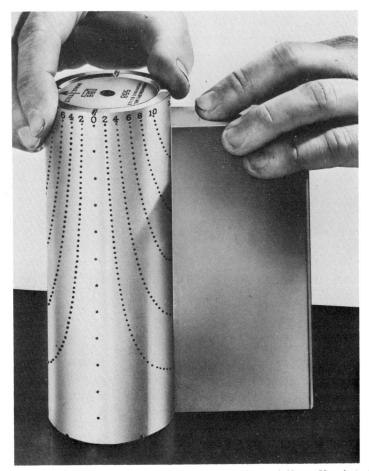

Courtesy of the Brown & Sharpe Manufacturing Co.

Fig. 2-11. Cylindrical square used to make precise measurements at right angles.

Vernier Measuring Instruments

The vernier is an auxiliary scale that is attached to calipers, height gages, depth gages, protractors, and other measuring instruments. It allows exceedingly accurate measurements to be made. The vernier measuring instrument is, in effect, a graduated steel rule to which a sliding member containing the vernier scale is attached. There are two types of vernier scales commonly used in the machine shop; one has a 25-division vernier scale and the other has a 50-division vernier scale.

The primary scale of a 25-division vernier is divided into inch graduations (large numbers), .100-inch graduations (small numbers), and .025-inch graduations (no numbers). The vernier scale has 25 divisions, and its length is equal to the length of 24 small divisions on the primary

Fig. 2-12. 25-division vernier scale.

scale. Thus, the length of the vernier scale is 24 × .025 or .600 inch. Since the vernier scale has 25 equally spaced divisions, the length of each division must be .600 ÷ 25, or .024 inch. Therefore, each division on the vernier scale is .001 inch (.025 − .024) shorter than each small division on the primary scale. When the zero graduations on both scales are aligned with each other, the first two lines on the primary and the vernier scales are .001 inch apart; the second two lines will be .002 inch apart; the third pair are .003 inch apart; etc. It is evident that in order to move the vernier exactly .002 inch from this position, it is only necessary to slide it along the primary scale until the second line on the vernier coincides with the corresponding line on the primary scale. To read a 25-division vernier follow the steps given below and refer to Fig. 2-12:

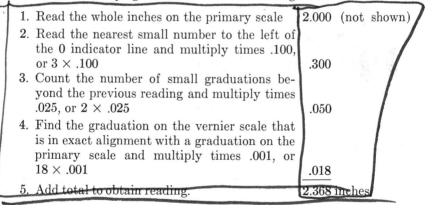

The 50-division vernier, Fig. 2-13, has more widely spaced graduations which are easier to read. The primary scale is divided into inch graduations, .100-inch graduations, and .050-inch graduations. The vernier scale is divided into 50 equally spaced graduations and its length is equal to the length of 49 divisions on the primary scale, or 49 × .050 = 2.450 inches. Each graduation on the vernier scale is then 2.450 ÷ 50, or .049 inch long. Thus, the difference between the length of a graduation on the primary scale and on the vernier scale is again .001 inch. To read a 50-division vernier follow the steps below and refer to Fig. 2-13:

24 BASIC MEASURING INSTRUMENTS Ch. 2

Courtesy of The L. S. Starrett Company

Fig. 2-13. 50-division vernier scale.

1. Read the whole inches on the primary scale 1.000
2. Read the nearest small number to the left of the 0 indicator line and multiply times .100, or 4 × .100 .400

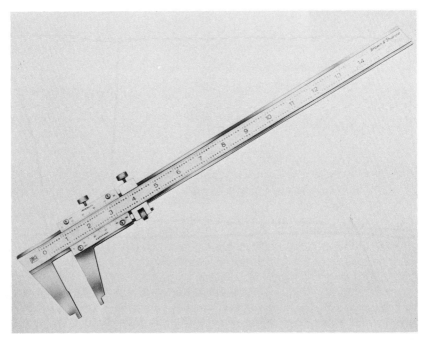

Courtesy of the Brown & Sharpe Manufacturing Co.

Fig. 2-14. Vernier caliper.

Ch. 2 BASIC MEASURING INSTRUMENTS 25

3. Count the number of small graduations beyond the previous reading and multiply times .050, or 1 × .050 .050
4. Find the graduation on the vernier scale that is in exact alignment with a graduation on the primary scale and multiply times .001, or 14 × .001 .014
5. Add total to obtain reading. 1.464 inches

Vernier calipers capable of measuring both inside and outside diameters are shown in Fig. 2-14. Vernier height gages (Fig. 2-15A, and Fig. 2-26) are used to measure vertical distances above a reference plane; usually a precision-surface plate. They are also used to scribe lines at a given distance above a reference plane when making precise layouts on workpieces. A vernier depth-gage is shown at B in Fig. 2-15.

Fig. 2-15. A. Vernier height-gage. B. Vernier depth-gage.

Vernier bevel protractors, as shown in Fig. 2-16, are used to measure angles to an accuracy of five minutes (05'). Since there are sixty minutes (60') in one degree, this is equal to one-twelfth ($5/60 = 1/12$) of a degree. An enlargement of the protractor scale and vernier is shown in Fig. 2-17. The main protractor scale is divided into degrees, with every ten degrees numbered. The vernier scale is actually *two* vernier scales, each having twelve divisions on either side of the zero graduation. The left-hand scale is used when the vernier zero graduation is moved to the left of the zero on the primary scale, while the right-hand scale is used when the movement is to the right. The total length of each vernier scale is equal to 23 degrees on the primary scale. Since there are twelve divisions on the vernier scale the length of each graduation is equal to $23 \div 12$ or $1\text{-}11/12$ degrees. Since one degree equals 60', $11/12$ degree equals $11/12 \times 60$ or 55'. Thus, the length of each graduation on the vernier scale is equal to

Courtesy of The L. S. Starrett Company

Fig. 2-16. Vernier bevel protractor.

Courtesy of The L. S. Starrett Company

Fig. 2-17. Vernier bevel protractor scale.

Ch. 2 BASIC MEASURING INSTRUMENTS 27

60′ + 55′ or 115′. One graduation on the vernier scale is, then, 5′ less in length than two graduations on the primary scale; two graduations on the vernier are 10′ less than four graduations on the primary; etc.

To read the vernier bevel protractor, first read the number of whole degrees passed by the vernier zero and then count, in the same direction, the number of graduations between the vernier zero and that line which exactly coincides with a graduation on the primary or degree scale; this number multiplied by 5 will give the number of minutes to be added to the whole number of degrees. In Fig. 2-17, the vernier zero has passed to the left of 50°, and the fourth line to the left of the vernier zero coincides with a line on the degree scale. Hence, the reading is 50° + 4 × 5′ or 50°20′.

Micrometer Measuring Instruments

The micrometer screw principle is used on a variety of measuring instruments which includes outside micrometer calipers, inside micrometer calipers, and micrometer depth gages. The precision obtainable by all micrometer measuring instruments is dependent upon the micrometer screw; therefore, the thread of this screw is made with the greatest possible care and degree of precision. For this reason, micrometer screw threads are very seldom made longer than one inch and micrometer measuring

Courtesy of The L. S. Starrett Company

Fig. 2-18. Outside micrometer caliper set including length standards.

instruments are designed to accommodate a one-inch movement of the micrometer spindle. For example, outside micrometer calipers are made with frames having sizes of one-inch increments. Figure 2-18 shows three outside micrometer calipers of the following sizes: 0 to 1 inch, 1 to 2 inches, and 2 to 3 inches. Inside micrometer calipers and depth micrometer calipers have separate stems for each one-inch interval within the range of the instrument.

Figure 2-19 illustrates a sectional view of a one-inch, outside micrometer caliper with all of the parts named. The principal parts are the frame, sleeve, thimble, spindle, and anvil. The frame supports the components of the micrometer caliper and keeps them in correct alignment.

Courtesy of The L. S. Starrett Company

Fig. 2-19. Sectioned view of 0-1 inch outside-micrometer caliper.

The sleeve houses the spindle nut and it has graduations marked lengthwise on its outside surface. These graduations are one inch long, and they are divided into .100-inch and .025-inch intervals. The thimble is attached to the spindle by a taper and a screw (see ratchet body in Fig. 2-19). The thimble rotates and moves lengthwise over the sleeve. One end of the thimble is beveled. The bevel on the thimble has twenty-five equally spaced graduations around its circumference. The spindle has the micrometer thread on one end and one of the two measuring surfaces on the face of the other. The other measuring surface is on the face of the anvil. These two measuring surfaces are made to be exactly parallel to each other regardless of the position of the spindle.

Ch. 2 BASIC MEASURING INSTRUMENTS 29

Fig. 2-20. Micrometer graduations.

The micrometer screw thread has 40 threads per inch. Its lead, or distance that the thread advances per revolution, is equal to 1 ÷ 40 or .025 inch or the distance between the smallest graduations on the sleeve. The beveled edge of the thimble will move lengthwise one graduation on the sleeve for each revolution, since it is attached to the screw. Turning the thimble through one graduation on the thimble scale will cause the thimble and spindle assembly to rotate exactly $\frac{1}{25}$ of a revolution, since there are 25 equally spaced graduations around the thimble. When the spindle rotates $\frac{1}{25}$ of a turn, it moves lengthwise $\frac{1}{25}$ of its lead, or $\frac{1}{25} \times .025 = .001$ inch.

To read a micrometer scale follow the steps below and refer to A in Fig. 2-20.

Courtesy of The L. S. Starrett Company

Fig. 2-21. Measuring part held in a lathe using outside-micrometer calipers.

1. Read the largest number visible on the sleeve scale and multiply by .100, or 2 × .100 .200
2. Count the number of small graduations beyond the graduation corresponding to the number read above, and multiply by .025, or 3 × .025 .075
3. Read the graduation on the thimble scale and multiply by .001, or 10 × .001 .010
4. Add to obtain reading. .285 inch

Some micrometers have a vernier scale, V, marked on the sleeve as shown at B, Fig. 2-20. The vernier scale, in conjunction with the regular micrometer scale, enables the micrometer to be read to .0001 inch. The relation between the graduations on the vernier and those of the regular micrometer is more clearly shown in C, in Fig. 2-20. The vernier has ten divisions which are the same length as nine graduations on the thimble. For convenience in reading, each graduation on the vernier scale is num-

Courtesy of The L. S. Starrett Company

Fig. 2-22. Measuring with large outside-micrometer caliper.

Courtesy of The L. S. Starrett Company

Fig. 2-23. The use of inside-micrometer calipers in measuring a large bore.

bered. The difference in width of a thimble division and a vernier division is equal to one-tenth of the thimble division. Therefore, a movement of the thimble equal to this difference results in a movement equal to $\frac{1}{10}$ of $\frac{1}{25}$ of a revolution, or $\frac{1}{10} \times \frac{1}{25} = \frac{1}{250}$ of a turn of the thimble and spindle assembly. This will cause the spindle to move lengthwise $\frac{1}{250} \times .025$, or .0001 inch.

In order to read a micrometer having a ten-thousandths vernier scale first determine the reading in thousandths, as with an ordinary micrometer, and then find a line on the vernier scale that exactly coincides with one on the thimble; the number on this line represents the number of ten-thousandths of an inch to be added to the number of thousandths obtained by the regular micrometer scales. For example, at C, in Fig. 2-20,

the micrometer scales read .275 inch. On the vernier scale, the fourth graduation coincides with a graduation on the thimble; therefore, 4 × .0001, or .0004 inch is added to the regular micrometer reading. The final reading is, then, .275 + .0004, or .2754 inch.

Accurate measurements made with outside micrometer calipers depend upon the way in which they are handled. In order to feel when they are in correct contact with the workpiece they must be held correctly when taking different types of measurements. Any movement of the frame, while manipulating the micrometer screw, should be kept to a minimum —just enough to establish the right feel. Methods of holding the micrometer are best shown by illustrations as follows: 1. Fig. 2-21, when a part is held in a lathe; 2. Fig. 2-7 and A, in Fig. 2-6, when the tool or part to be measured is held in the hand; 3. Figs. 2-1 and 2-22, when measuring with large micrometer calipers. Inside micrometer calipers, as shown in Fig. 2-23, must be manipulated through a hole, or recess, much in the same manner as inside calipers, in order to find the true diameter of the hole. It is advisable to check the accuracy of inside micrometer calipers with outside micrometer calipers before they are used. The use of depth micrometers is shown in Fig. 2-24. The thimble is carefully turned until the measuring rod just touches the surface to be measured.

Courtesy of The L. S. Starrett Company

Fig. 2-24. Using depth micrometers to measure shoulder depth.

Ch. 2 BASIC MEASURING INSTRUMENTS 33

Courtesy of The L. S. Starrett Company

Fig. 2-25. Dial test indicator used to check the size of a machined surface on a planer with a planer gage shown in the background.

Dial Test Indicators

The dial test indicator (see Fig. 2-25) is a sensitive instrument having a graduated dial and an indicating hand which is connected through a system of multiplying levers, or gears, to a contact point that is fastened to the end of a spindle. The contact point is placed in contact with the workpiece. Any movement of the contact point is transmitted through the spindle, greatly amplified by the internal works, and displayed by a movement of the indicating hand on the dial. Dial test indicators are designed to read to .001 inch, .0005 inch, or .0001 inch. The reading of the divisions is usually marked on the dial face. They are made in a great variety of sizes and shapes. Indicators are widely used in machine shops, and tool and die shops in combination with many other forms of gaging and measuring devices. In the shop, dial test indicators are frequently simply referred to as "indicators."

The dial test indicator in Fig. 2-25 is used to check the depth of cut on a flat casting. The indicator reading is compared to the setting of a planer gage shown in the background. Dial test indicators are used with vernier height gages, Fig. 2-26, to assure uniform contact pressure. The height gage is set by making the indicator read zero on all surfaces to be measured. Among other applications of indicators are these: to locate edges and holes for precision boring, to measure concentricity, to "zero"

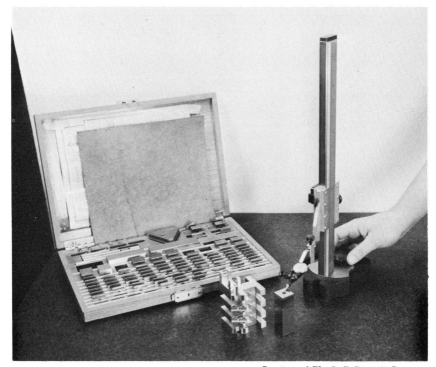

Courtesy of The L. S. Starrett Company

Fig. 2-26. Precision gage-blocks used with dial test indicator attached to vernier height-gage to measure workpiece.

the position of machine tool slides, and to measure parallelism between surfaces. The operation of using a dial test indicator is usually called "indicating."

Precision Gage Blocks

Precision gage blocks, Fig. 2-26, are small blocks of steel that have been heat treated to a high hardness. The heat treatment is also designed to obtain a high degree of dimensional stability in the blocks. They are made to have an extremely high degree of dimensional accuracy as shown in Table 2-1. There are several grades of gage blocks which can be categorized as master blocks, inspection blocks, and working blocks. Working blocks are used in the workshop. They are used to calibrate micrometers and other precision measuring tools, to inspect dies, tools, and other precision machine parts. In Fig. 2-16 a gage block is shown on a precision-surface plate. It is used with the height gage and the dial test indicator to measure the height of a surface on the workpiece by comparing the height of the gage block and the workpiece with the indicator. Gage blocks are also used to measure angles in conjunction with sine bars and

Table 2-1. Accuracy Standards for Precision Gage Blocks

Classification	English Measure Standards			Surface Finish	Metric Measure Standards		
	Length	Flatness	Parallelism		Length	Flatness	Parallelism
	inches	inches	inches	microinches	millimeters	millimeters	millimeters
U.S. Federal AA	±.000002	.000004	.000003	.5	±.00005
British Reference Grade	±.000002	.000003	.000003	...	±.00005	.00008	.00008
U.S. Federal A	+.000006 / −.000002	.000004	.000004	.8	+.00015 / −.00005
British Calibration Grade	±.000005	.000003	.000003	...	±.00012	.00008	.00008
German Grade O	+.000010 / −.000006	+.00025 / −.00015	.00010	.00010
U.S. Federal B	+.000007 / −.000003	.000006	.000005	1.2	+.00020 / −.00010
British Inspection Grade000005	.000005	...	±.000030	.00010	.00010
German Grade I00015	.00015
U.S. Federal
British Workshop Grade	+.000010 / −.000005	.000010	.000010	...	+.00025 / −.00010	.00025	.00025
German Grade II	±.00070	.00025	.00025
German Grade III	±.00140	.00050	.00050

U.S. Federal Specification GGG-G-15.—British Standards 888-1950.—German Standards DIN 861.

Courtesy of the Brown & Sharpe Manufacturing Co.

Fig. 2-27. Precision gage-blocks and sine plate used to measure angles.

sine plates, as shown in Fig. 2-27. Angles are sometimes machined on sine plates which are set by using gage blocks. Precision, angular gage blocks are also made for making angular measurements.

Gage blocks are sold in sets. A set of gage blocks is shown in Fig. 2-26. The sizes for an 83-piece, gage-block set are given in Table 2-2. By combining two or more gage blocks, a large range of extremely accurate dimensions can be obtained. The blocks can be combined in the following way when there is in the set no single block of the exact size that is wanted. Two or more blocks are assembled, or combined, by wringing the blocks together to form the equivalent of a single block of a given dimension. First, the joining surfaces of the blocks are cleaned. One block is then placed over the other at an angle and the two are twisted, or wrung, together. When properly wrung together, the blocks will adhere to each other so strongly that a very considerable force is required to separate them. They are best separated by twisting them apart in a manner similar to the way in which they were assembled.

In selecting gage blocks for a given dimension, the least number of blocks that will give the required dimension should be used. This is accomplished by successively eliminating the smallest remaining dimension.

Example 2-1.

Determine the gage blocks required to obtain **3.6742** inches. (Refer to Table 2-2.)

Table 2-2. Sizes for an 83-piece Gage-Block Set

First: .0001 Series (9 Blocks)

.1001	.1002	.1003	.1004	.1005	.1006	.1007	.1008	.1009

Second: .001 Series (49 Blocks)

.101	.102	.103	.104	.105	.106	.107	.108	.109	.110	.111	.112	.113
.114	.115	.116	.117	.118	.119	.120	.121	.122	.123	.124	.125	.126
.127	.128	.129	.130	.131	.132	.133	.134	.135	.136	.137	.138	.139
.140	.141	.142	.143	.144	.145	.146	.147	.148	.149			

Third: .050 Series (19 Blocks)

.050	.100	.150	.200	.250	.300	.350	.400	.450	.500	.550	.600
.650	.700	.750	.800	.850	.900	.950					

Fourth: 1.000 Series (4 Blocks)

1.000	2.000	3.000	4.000

Two .050 Wear Blocks

Step 1. Eliminate the .0002 inch by selecting a .1002 block. Subtract .1002 from 3.6742.

$$\begin{array}{r} 3.6742 \\ -.1002 \\ \hline 3.5740 \end{array}$$

Step 2. Eliminate the .004 inch by selecting a .124 block. Subtract .124 from the remainder in Step 1.

$$\begin{array}{r} 3.574 \\ -.124 \\ \hline 3.450 \end{array}$$

Step 3. Eliminate the .450 inch with a .450 block. Subtract .450 from the remainder in Step 2. This obviously leaves 3.000 which is obtained with a 3-inch block.

Step 4. Check the blocks to determine if they will result in the required dimension.

$$\begin{array}{r} .1002 \\ .124 \\ .450 \\ 3.000 \\ \hline 3.6743 \text{ inches} \end{array}$$

Surface Gage

The surface gage, Fig. 2-28, is one of the most useful tools employed by the machinist and the toolmaker. On certain classes of work it is almost an indispensable tool. For example, surface gages can be used to scribe layout lines on workpieces in order to locate the centers of holes, to provide reference lines for setting up workpieces on the machine, or to outline surfaces to be machined. In Fig. 2-28, the scriber point of the surface gage is being positioned by a rule in preparation for scribing a line on the workpiece. The rule is conveniently held in a vertical position by a rule holder; a surface plate provides a reference plane for transferring the measurement from the rule holder to the workpiece by means of the surface gage. Located at the end of a rocker plate, a vertical thumbscrew provides a very fine adjustment to the spindle and allows accurate settings of the scriber to be made.

A dial test indicator, which can be used to transfer precise measurements, can be attached to the spindle or to the scriber. The scriber point can be used to check the position of layout lines when setting up a workpiece on the machine. The bent point of the scriber can be used to lay out lines that are close to the base of the surface gage. It is also used to test the parallelism between a surface on the workpiece and the top of a surface plate or machine tool table. This can be done whether the surface of the workpiece has been finished smooth or if it is still rough. The use

Courtesy of The L. S. Starrett Company

Fig. 2-28. Surface gage used to transfer a measurement from a 12-inch rule. The rule is held in a rule holder.

of the surface gage to test a cast surface for parallelism is especially effective because the surface gage can average out the irregularities on such surfaces.

Universal Precision Gage (Planer Gage)

A universal precision gage is shown at left in Fig. 2-29. This gage was developed from a planer gage by the incorporation of a fine adjustment mechanism. Originally, planer gages were used for setting planer and shaper cutting tools to establish the correct depth of cut. The universal precision gage can still be used for this purpose. Many other uses have been found for this tool in doing precision machine-tool work. The slide can be precisely adjusted to any vertical position by measuring across the base and the top of the slide with a micrometer caliper, or by transferring the measurement from a stack of gage blocks with a height gage

Courtesy of The L. S. Starrett Company

Fig. 2-29. (Left) Setting a universal precision gage using a height gage, dial test indicator, and precision gage-blocks. (Right) Scribing a layout line with a universal precision gage.

and a dial test indicator, as shown in the illustration at right. A scriber can then be attached to the universal precision gage in order to scribe layout lines. Other uses include measurements of parallel faces in a slot, as an adjustable parallel, for setting up and leveling workpieces, and as a height gage for transferring measurements from gage blocks by using a dial test indicator mounted on the universal precision gage.

The Basic Nomenclature of Measurement

The nomenclature of measurement deals with the definitions of the terms encountered in the field of measurement. The topics in the remainder of this book will be more meaningful after a study of these terms.

Nominal Size is the ideal size to which a part, or two mating parts, is to be made. It does not take into account the clearance, or interference necessary in fitting two parts together, so that they will function as desired. Nor does it take into account the deviations in size resulting from the manufacturing process.

Allowance is the intentional clearance between two mating parts so that they will work together as desired when they are assembled. In the case of a press, or shrink fit, allowance is the intentional interference between two parts.

Limits are the maximum and the minimum permissible dimensions of the size of a part. For example, a bore may be specified to have a diameter between 6.750 and 6.756 inches, in which case 6.750 and 6.756 inches are the limits.

Upper Limit specifies the largest size to which a dimension on a part is to be made. In the example above, 6.756 inches is the upper limit.

Lower Limit specifies the smallest size to which a dimension on a part is to be made. In the above example, 6.750 inches is the lower limit.

Ch. 2 BASIC MEASURING INSTRUMENTS 41

Tolerance is the specification of the total permissible variation of size of a dimension on a part. The total tolerance in the above example is .006 inch. In some instances this is expressed in terms of plus or minus from a given basic size (basic size is described below). For example, the tolerance in the example above might also be expressed as ±.003.

Basic Size is that size from which the limits of a part are obtained by the application of tolerances. It is sometimes considered as the "ideal" size of the part. For example, if the ideal size or the basic size of a bore is 6.753 inches and the tolerance is ±.003, the limits of the hole would be 6.750 and 6.756 inches.

Unilateral Tolerance refers to a permissible deviation in the basic size in one direction only; i.e., this specification states that the part is to be made with the basic size as the lower limit, or with the basic size as the upper limit. An example of a unilateral tolerance would be 6.750 + .003. Another example would be 6.775 − .002.

Bilateral Tolerance refers to a permissible deviation from the basic size in two directions; i.e., this specification would state that the basic size is neither the upper nor the lower limit and that the part may be slightly larger or slightly smaller than the basic size. Bilateral tolerances may be equal or unequal. An example of an equal bilateral tolerance is 6.753 ±.003 inches. An example of an unequal bilateral tolerance would be $6.752 \, ^{+.002}_{-.001}$ inches.

Precision refers to the degree of dimensional refinement to which a part is made or to which a machine is capable of working. A part which is made to, or a machine which will produce parts, with a very small dimensional tolerance is said to be precise.

Accuracy refers to the confirmation of a part to the requirements specified by the dimensional tolerance. A part that is within the required dimensional tolerance is accurate. If the required tolerance is very small, say ±.001 inch, the part is also precise. If the tolerance on a machined part is large, say, ±1/16 inch, a part is accurate if it falls within this range, yet, at the same time, it would be considered to be imprecise.

QUESTIONS

2-1. Determine the decimal equivalent of each of the following fractional inch dimensions:

 a. 7/8 d. 27/32 g. 3/64
 b. 5/8 e. 27/64 h. 19/32
 c. 11/16 f. 1 9/16 i. 2 45/64

2-2. Describe two methods of testing the accuracy of a square.

2-3. Make a sketch of a 25-division vernier scale which is set to each of the following readings:

| | a. 4.527 inch | c. 3.929 inch |
| | b. .777 inch | d. 1.042 inch |

2-4. Make a sketch of a 50-division vernier scale which is set to each of the following readings:

| | a. 3.272 inch | c. 8.069 inch |
| | b. .467 inch | d. 1.191 inch |

2-5. Make a sketch of a micrometer scale which is set to each of the following readings:

	a. .456 inch	d. .888 inch	g. .120 inch
	b. .782 inch	e. .312 inch	h. .981 inch
	c. .042 inch	f. .671 inch	i. .501 inch

2-6. Make a sketch of a micrometer scale with an auxiliary vernier scale which is set to each of the following readings:

	a. .2422 inch	e. .0043 inch
	b. .9685 inch	f. .5932 inch
	c. .5128 inch	g. .3734 inch
	d. .6246 inch	h. .1402 inch

2-7. Make a sketch of the vernier bevel protractor scale set to each of the following readings:

| | a. 24°25′ | c. 10°05′ |
| | b. 73°18′ | d. 32°43′ |

2-8. Select the sizes of the precision gage blocks required to obtain each of the following lengths of block assemblies. Assume that the gage blocks listed in Table 2-2 are available.

	a. .6721 inch	f. 3.7513 inch
	b. 12.2442 inch	g. 9.4375 inch
	c. 8.4529 inch	h. 1.0001 inch
	d. 2.0055 inch	i. .3258 inch
	e. 3.7513 inch	

2-9. What is the difference in meaning between the words accuracy and precision?

2-10. What is the nominal size of a part?

2-11. What is the difference in meaning between allowance and tolerance?

2-12. What are the limits of a dimension?

2-13. Name the tolerance, upper limit, and the lower limit of each of the following dimensions.

| | a. 6.788 ±.005 | c. 1.000 $^{+.0005}_{-.0000}$ | e. .499 / .496 |
| | b. .542 $^{+.000}_{-.001}$ | d. 3.062 $^{+.002}_{-.003}$ | f. 1.625 / 1.620 |

2-14. A dimension on a part must not exceed .250 inch or be less than .248 inch. Specify the dimension of this part by each of the following methods:

 a. Bilateral tolerances
 b. Unilateral tolerances
 c. Limit dimensions.

CHAPTER 3

Layout Work

Castings, forgings, weldments, and rolled sections are laid out in order to establish reference lines or planes which are used to set up the part on the machine, to outline the surfaces that are to be machined, and to show the approximate amount of metal to be removed from each surface. Castings and forgings in particular, have excess material on all surfaces which are to be machined, including cored holes. One purpose of the layout is to determine if enough metal exists on all of these surfaces to "clean up" or "machine out." Sometimes, a core will shift in making a casting, resulting in insufficient metal to "machine out." When the core shift is only a small amount this can be corrected in the layout; at other times, the casting must be scrapped. However, it is best to know that the casting will be scrap *before* any expense is encountered in machining.

Two other reasons for laying out workpieces that are to be machined are to locate and outline the position of holes that are to be drilled and to outline keyways that are to be cut. Although castings, forgings, and weldments are usually laid out before machining any surface, additional layout work is often done after one or more machining operations have been finished. In some cases, usually involving rolled material, the initial layout operation is performed after several machining operations are completed. Not all workpieces are laid out. Parts made from bar stock and those parts having a relatively simple shape are often machined directly, without making a layout.

The layout is usually made from a reference surface which is a plane. Surface plates, layout plates, layout tables, floor plates, and the top of machine-tool tables are used in making layouts. They must be kept clean and free from scratches, nicks, and burrs that would impair the accuracy of the reference surface and the layout work done upon it. Other tools used to do layout work include precision steel rules, layout rules, squares, bevel protractors, surface gages, height gages, scribers, hermaphrodite calipers, dividers, prick punches, center punches, hammers, V-blocks, parallels, angle plates, and a variety of clamps. Layout machines (see Figs. 3-13 and 3-14) have been developed that greatly facilitate the layout operation.

Painting for Layout

The first step in making any layout is to paint the surface. Surfaces that have a smooth finish are usually painted with a blue layout dye.

Ch. 3 LAYOUT WORK 45

Brushed onto a clean surface it dries rapidly, providing a blue background that makes scribed lines stand out sharply and clearly. The surfaces of rough castings and forgings are usually painted with a specially prepared white paint. This paint is composed of a suitable white pigment mixed with alcohol, which evaporates rapidly upon application, enabling the paint to dry in a short time. This mixture should be thin so that it will spread easily on the metal surfaces. White blackboard chalk can be used when no paint is available; however, it is not permanent and is not a good substitute for layout paint or dye.

Layout of Hole Location for Drilling

The procedure for laying out holes that are to be drilled in a thick plate is shown in Fig. 3-1, A through F. The dimensions of the hole locations are given in F. The surface finish of the plate is smooth; free from nicks and burrs. After painting the surface that is to be laid out, the workpiece is clamped to the angle plate with one edge resting on the layout table as in A. For clarity, the clamps are not shown. Using a precision steel rule that is held in an upright position, the surface gage, shown in A, is set so that the scriber point is exactly 5 inches (1¾" + 3¼") above the lower edge of the workpiece. With this setting the surface gage is then used to scribe a horizontal line as shown in B. In a similar manner, a second horizontal line is scribed 1¾ inches above the lower edge of the workpiece. With a steel rule positioned as shown in C, the

Fig. 3-1. Steps in laying out holes for drilling.

1⅝-inch, 3⅜-inch, and the 4-inch distances are laid out with a scriber. The location of each hole is marked by two scribed lines which come together to a point. The two lower holes are laid off by scribing two vertical lines with a scriber and a square, as shown at D. An alternate method is illustrated in E, where the workpiece is turned 90 degrees and reclamped on the angle plate. The workpiece is correctly positioned by aligning the scribed line with the blade of a square. The 1⅝-inch, 3⅜-inch, and 4-inch dimensions are then laid out with a surface gage.

A prick punch is carefully positioned at the center of each hole, as determined by the layout lines, and a mark is made by tapping the punch with a hammer. Using a rule, dividers are then set to the radius of the hole to be drilled, and a circle, equal in diameter to the hole diameter, is scribed at each hole location. (See Fig. 3-2.) The center of each hole is punched again with a center punch and a hammer, and four "witness marks" are punched on the scribed circles as shown in F, Fig. 3-1. The purpose of the witness marks is to provide points of observation when starting to drill the hole with the point of the twist drill. The scribed lines can be rubbed off by the chips but the witness marks will remain. The circle and witness marks are usually omitted for holes that are approximately ¼ inch, or less, in diameter.

There are other procedures that can be used to lay out holes, two of which are shown in Fig. 3-3. At the left, a combination square and rule is used, while at the right, the layout is made with hermaphrodite calipers. Often, **combinations of all of these procedures are used to lay out a part.**

Courtesy of The L. S. Starrett Company

Fig. 3-2. Scribing hole diameters on a hole layout using spring dividers.

Layout of Cylindrical Part

Layout operations on cylindrical parts usually require that the point of the surface-gage scriber be set on the exact center of the workpiece. The procedure for doing this is shown in Fig. 3-4, A through F. This procedure is preliminary to finding the exact center of the end faces of the workpiece, prior to center drilling the ends in preparation for turning the part between centers in a lathe. (See D.) It is also required when holes and keyways are to be laid out, as shown in E and F.

Ch. 3 LAYOUT WORK 47

Courtesy of The L. S. Starrett Company

Fig. 3-3. (Left) Using combination square and scriber to lay out plate. (Right) Using hermaphrodite calipers to lay out plate.

Fig. 3-4. Laying out cylindrical work; A through C. Method of setting surface-gage scriber on centerline of work axis. D. Laying center of end face. E. Laying out for drilling hole perpendicular to axis. F. Laying out for drilling holes perpendicular to axis and at an angle to each other.

The part is first painted in the areas to be laid out. After the paint has dried, the part is placed in a V-block which is standing on a layout table. The procedure for laying out the cylindrical part for the different configurations mentioned above will be described in the following steps, and by referring to the diagrams in Fig. 3-4.

1. Adjust the position of the surface gage scriber to the estimated center of the work.
2. Scribe a line from A to B. (See diagram A, Fig. 3-4.)
3. Rotate the work approximately 180 degrees or one-half turn, until the B end of line AB is aligned with the surface-gage scriber.
4. Scribe line BA'. (See diagram B, Fig. 3-4.)
 Note: $A'B$ may come out above AB, instead of below AB, as shown.
5. Adjust the surface-gage scriber one-fourth the distance from A' to A. (If $A'A$ coincides with $A'B$ as in diagram C, Fig. 3-4, the surface-gage scriber is set at the exact center of the work, thus steps 5 and 6 are not required.)
6. Repeat steps 2 through 5 until AB coincides with $A'B$ as shown in diagram C. Two or three trials should be all that are normally required to obtain this setting of the surface-gage scriber.
7a. To locate the center of the end face for drill and countersinking in preparation for turning between centers in a lathe, scribe two additional lines with the surface gage as shown in D, Fig. 3-4. The intersection of any two of these lines will be the exact center. The third line serves as a check on the accuracy of the work. If all three lines intersect at the same point, the work is accurate.
7b. The layout at E, Fig. 3-4 is for a hole that is to be drilled both perpendicular to, and through, the axis of the work. The scribed line is extended along the cylindrical surface of the work with the surface gage scriber set "on center." The distance from the end face to the center of the hole is laid out with hermaphrodite calipers, as shown, or with a scriber and a rule. This location is prick punched, and a circle, equal to the hole diameter, is scribed with dividers. The center of the hole is center punched and witness marks are punched around the circle. Often, two very light prick punch marks are made on the scribed line on the end face, as shown, which serves as a guide when setting up the part on the drill press.
7c. The procedure for laying out a keyway at an angle to a hole is illustrated in F, Fig. 3-4. The hole is laid out as explained in step 7b. A bevel protractor is then set to do the required angle. The scribed line on the end face corresponding to the hole is then aligned with the blade of the protractor, as shown. Thus, the work is positioned in the V-block for laying out the keyway. A

line is scribed across the end face and along the cylindrical portion of the work with the surface gage set "on center." Using a rule, the scriber of the surface gage is set above center; a distance equal to one-half of the width of the keyway. A line is then scribed with the surface gage as shown in Fig. 3-4, F. Similarly, a line is scribed below the center to outline the sides of the keyway. The depth of the keyway is marked, and using a scriber and a square, a vertical line is scribed on the end face. The length of the keyway is laid out with a scriber and a rule. A scriber mark and a prick punch mark are then made at a distance equal to one-half of the width of the keyway, from the end. Using dividers set in the prick-punch mark, a half circle is then scribed at the end of the keyway layout to complete the job.

Procedures other than those described in step 7a can also be used to locate the center of the end face. When the cylindrical surface is rough, the use of the hermaphrodite calipers will tend to give more accurate results than do other methods as the caliper leg will contact a greater length of the cylindrical surface as it is moved along, thereby averaging out small irregularities. The method of using hermaphrodite calipers is shown in A, Fig. 3-5. When the cylindrical surface is fairly true, a center head, shown in Fig. 3-5, B, is a fast and accurate method of performing this operation. Two or three lines are scribed with the blade held in different positions by a scriber. The intersection of these lines is the center of the end face. A center punch mark C, is made to mark the center. In Fig. 3-5, D, a bell center is shown. The conical surface locates the punch at the center of the end face without the necessity of making a layout. The punch is struck with a hammer, to mark the center. Bell centers should only be used when the workpiece is fairly round.

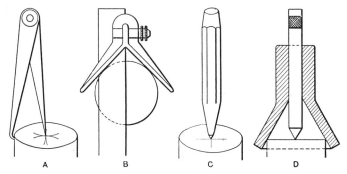

Fig. 3-5. Laying out center, on end face of cylindrical work. A. Using hermaphrodite calipers; B. Using center head; C. Using center punch; D. Using bell center.

Layout of Equally Spaced Holes on Hole Circle

A form of construction frequently encountered is the placement of holes in a circular pattern, as shown in Fig. 3-6. The circle on which the

hole is located is called a *hole circle*. The center of the hole circle is often located in the center of a larger central hole. In such a case, it will be necessary to press a wooden plug or stick (see Fig. 3-6) into the central hole. Prior to pressing the plug or the stick into the hole, a large-head nail is driven into the center of the stick. The center of the central hole can now be found with hermaphrodite calipers, or by using a surface gage and a rule. A light prick-punch mark made at the intersection of the scribed lines on the nail head locates the center of the central hole. This prick-punch mark anchors one leg of the dividers, or trammel points, while the other leg is swung in an arc to scribe the hole circle.

The next operation is to set the dividers to a distance D (see Fig. 3-6), and to scribe the center of each hole. Usually, the position of one of the holes is defined by the drawing; however, in some cases, no starting location is given, as in the case of the ring in Fig. 3-6. In either case, a prick-punch mark is made at the center of the first hole. The dividers, set at the distance D, are then positioned in this prick-punch mark and a short arc is scribed across the hole circle. A prick-punch mark is made at the intersection of the arc and the hole circle. This procedure is repeated until the centers of all of the holes have been located. To complete the layout, circles are scribed with dividers and witness marks are punched.

One of the problems that can be time consuming is the determination of the distance D, at which the dividers must be set to obtain equally spaced holes around the hole circle. A unique situation arises when the number of equally spaced holes is six. In this case, the distance D is equal to the radius of the hole circle. When there are relatively few holes equally spaced around the hole, circle D can be determined by trial-and-error. When there are many holes, however, the trial-and-error method is slow

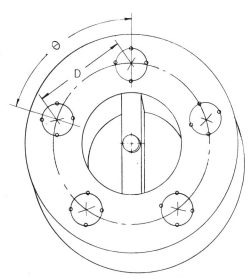

Fig. 3-6. Laying out holes on bolt circle.

and inaccurate. A better way is to calculate D, as explained in the two steps below (see Fig. 3-6):[1]

1. Calculate the included angle θ, which is the angle between two adjacent holes on the hole circle. This is done by dividing the number of holes in the hole circle into 360° or

$$\theta = \frac{360°}{\text{No. holes in hole circle}}$$

2. Calculate D by multiplying the diameter of the hole circle by the sine of one-half of the angle θ.[2]

$$D = \text{Diameter of hole circle} \times \sin \frac{\theta}{2}$$

Example 3-1:
Twenty equally spaced holes are to be placed around a 10-inch diameter hole circle. Calculate the distance between each adjacent hole.

1. $\theta = \dfrac{360°}{\text{No. holes in hole circle}}$

 $= \dfrac{360°}{20}$

 $= 18°$

2. (From a table of trigonometric functions, $\sin 9° = .15643$)

 $D = \text{Diameter of hole circle} \times \sin \dfrac{\theta}{2}$

 $= 10 \times \sin \dfrac{18°}{2}$

 $= 10 \times \sin 9°$

 $= 10 \times .15643$

 $= 1.5643$ inches

This is approximately $1\frac{9}{16}$ inches, since the decimal equivalent for $\frac{9}{16}$ inch is .5625 inch.

Layout of Large Castings

Large castings vary so greatly that the layout procedure to be described must be rather general. There are, however, certain specific steps to be taken when laying out any large casting which, when completed, accomplish a desired objective. These steps, given below, can serve as a guide for laying out any large casting:

[1] A table of chord lengths for spacing off circles for a wide range of hole numbers is given in Machinery's Handbook, Industrial Press Inc.

[2] The sine can be found in any table of natural trigonometric functions that are given in various handbooks.

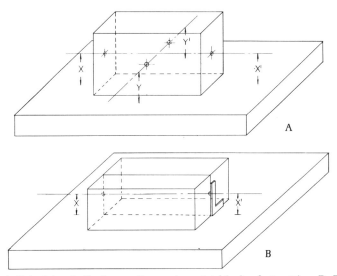

Fig. 3-7. A. Principle of aligning casting on layout table for first setting. B. Principle of aligning casting on layout table for second setting.

1. Place the work on parallel blocks that have been previously placed in position upon the layout table. Since the surface of the casting is rough, it is usually necessary to add shims to prevent the casting from rocking. Screw jacks are used, sometimes, in place of parallel blocks and shims.
2. Paint the casting with a white layout paint on all surfaces on which layout lines are to be scribed.

Courtesy of the Giddings & Lewis Machine Tool Company

Fig. 3-8. Checking starting reference point of layout with a surface-gage and a ruler.

3. Establish a reference plane by selecting and leveling two horizontal lines, or planes, that are at right angles to each other on the casting. This is the initial step in actually making the layout. The principle of establishing a reference plane is illustrated in A, Fig. 3-7. A block having two perpendicular axes is shown. Each axis is parallel to the top and the bottom surfaces, or planes, on the block. The pair of axes may or may not actually intersect each other. The top and bottom surfaces, or planes, of the block will be parallel to the top of the layout table when the distances Y and Y', taken from the top of the table, are equal to each other and when the distances X and X' are equal to each other. It is not necessary for Y and Y' to be equal to X and X'.

In order to establish the reference plane on the casting, a starting point must be selected. Sometimes, a flat surface can be used as a reference plane, in which case this surface is leveled with a surface gage or placed in a perpendicular position with a square.

Courtesy of the Giddings & Lewis Machine Tool Company

Fig. 3-9. Setting the layout scale to the starting reference point.

Care should be exercised in selecting plane surfaces as reference planes because they often are made to have "draft," an intentional taper which is provided to enable the pattern to be removed from the mold in the foundry. Cored holes seldom are made with draft. The axis of a cored hole, therefore, frequently affords a better reference position than does a plane surface. The axes of two perpendicular cored holes also are frequently used to establish a reference plane on a casting. Figure 3-8 shows the center of a cored hole being checked with a rule and a surface gage. This cored hole passes through the casting and the axis of the cored hole is selected to establish the reference plane. The casting is leveled until the cored hole axis is parallel to the layout table; this is accomplished by making the distance from the table top to

Courtesy of The R. K. LeBlond Machine Tool Company

Fig. 3-10. Scribing a finish line for a surface to be milled with a surface gage.

the center of the cored hole, equal at each end. If necessary, shims are used to adjust the position of the casting. A similar procedure is used to find, and then to level the second axis, which is perpendicular to the first axis. When adjusting the casting to bring the second axis parallel to the layout table, raise and shim two corners of the casting an equal amount. This is necessary in order to maintain the parallelism of the first axis.

4. Position the rule, or if used, the layout scale, with respect to the casting. A layout scale, Fig. 3-9, has graduations which have been specially cut to mark the dimensions that are to be laid out on a particular casting model. When quantities of the same casting

models are to be made, the layout scale will pay for itself by reducing the time required to make the measurement and by reducing errors in reading the rule. When only one or two castings are to be laid out, a precision steel rule is used. In either case, the layout scale, or the rule, is adjusted vertically on the rule holder until one of its principal graduations (say, an inch mark) is exactly aligned with the point of the surface-gage scriber, with the scriber positioned on the reference axis. In Fig. 3-9, two lines are shown scribed across the cored hole containing the reference axis; the layout scale is being positioned with respect to this axis, using the surface gage.

5. Check the amount of metal available for machining on all surfaces that are to be machined. The purpose of this step is to determine

Courtesy of the Giddings & Lewis Machine Tool Company

Fig. 3-11. Prick punching the first setup line.

if enough metal exists on each surface in order to "machine out" or "clean up." The dimension of each surface is checked with the surface gage, then a line is scribed to mark the location of the surface to be machined. Figure 3-10 shows a finish line being scribed on a casting, to define a surface that is to be milled or planed.

The amount of metal to be machined in cored holes is checked by scribing two lines, one above and one below the axis of each cored hole. The same cored hole is checked again in the second setup of the casting when the casting is positioned 90° from its present position, as described in Step 7. Thus, the hole will have four scribed lines as shown in Figs. 3-13, 3-14, and 3-15, which will

determine if sufficient metal is available to clean up when it is machined.

The layout procedure sometimes will show that there is insufficient metal on a plane surface, or on a hole, to "clean up." In this event, the rule or the layout scale is moved up or down, as required, but far enough so that there will be sufficient metal to

Courtesy of the Giddings & Lewis Machine Tool Company

Fig. 3-12. Resetting work to first setup line using a square.

"clean up" the defective surface. Every other surface must now be checked again to make certain that this move will not result in the failure of some other surface to be cleaned up. Occasionally, a core will shift sufficiently in making a casting to cause the cast surfaces to be badly misaligned. In such a case the layout will indicate that it is impossible to "machine out" all surfaces.

Ch. 3 LAYOUT WORK 57

The casting must then be scrapped unless it can be salvaged by welding.
6. Scribe the first setup line around the casting. It should then be prick punched at convenient intervals, as shown in Fig. 3-11. The setup line defines a plane that is parallel to the two original mutually perpendicular axes that were selected in Step 3. This line will determine the alignment of the setup on the machine

Courtesy of the Giddings & Lewis Machine Tool Company

Fig. 3-13. Scribing the second setup line.

and it will also indicate a starting point from which other machined surfaces are dimensionally tied. It is sometimes convenient to have the setup line coincide with the center of the casting.
7. Establish the second reference plane. The second reference plane is required to check the amount of material on all surfaces for "clean up," and to scribe a second setup line which is used to align the part in the machine when setting up. The work is reset

on the layout table by turning it 90°. The principle of aligning the part on the layout table for the second setting is shown in B, Fig. 3-7. The first layout line (Step 6) is aligned perpendicular to the top of the layout table, using a square, as shown in B, Fig. 3-7 and also in Fig. 3-12. The original axis (the cored hole in the casting illustrated) is then made to be parallel to the top of the layout table by making distance X equal to X' (see B, Fig. 3-7). In

Courtesy of Portage Double Quick, Inc.

Fig. 3-14. Layout machine used to lay out casting.

practice, this is accomplished with a rule and a surface gage as shown in Fig. 3-8, and described in Step 3.

Repeat steps 4, 5, and 6. These steps are repeated in order to check the amount of metal available for machining on all surfaces to be machined, and to scribe the second setup line around the casting. Figure 3-13 shows the second setup line being scribed.

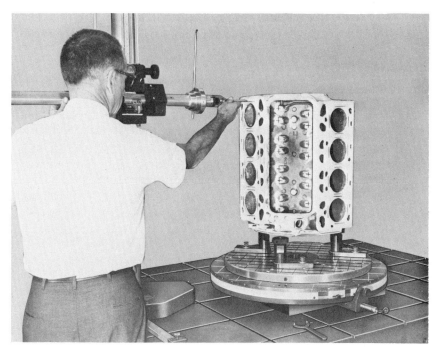

Courtesy of Portage Double Quick, Inc.

Fig. 3-15. Layout machine used to lay out V-8 cylinder block.

Layout Machines

The layout work just described can be greatly simplified, and the accuracy greatly increased, by the use of layout machines. A layout machine is illustrated in Fig. 3-14. The layout scriber can be moved horizontally, or parallel to the layout table, by moving the entire machine and by sliding the horizontal arm. It can be moved vertically by moving the arm assembly up or down the column. Vernier scales attached to the column and the arm make possible accurate vertical and horizontal movements. Circles can be scribed with the circular scribing attachment shown in Fig. 3-14.

One of the great advantages of the layout machine is that the need for a second setting is generally eliminated. All surfaces can be laid out in one setting. Figure 3-15 illustrates an automobile cylinder positioned on a rotary table and being laid out with a layout machine.

QUESTIONS

3-1. Name four purposes for making a layout on a part that is to be machined.

3-2. Name twelve tools commonly used to do layout work.

3-3. Why are parts painted prior to being laid out?

3-4. When is a blue layout dye used to paint the surface? When should a white layout paint be used?

3-5. What are "witness marks" and why are they used?

3-6. Describe how the center on the end face of a cylindrical part is laid out, using a V-block and a surface gage.

3-7. Nine equally spaced holes are to be laid out on a 15-inch hole circle. Calculate the divider setting required to obtain the correct distance between each adjacent hole.

3-8. Twenty equally spaced holes are to be laid out on a 24-inch hole circle. Calculate the divider setting required to obtain the correct distance between each adjacent hole.

3-9. List in their correct order the steps that are required to lay out a large casting.

3-10. What is the layout scale and how does it differ from a steel rule?

3-11. What must be done if a surface fails to "clean up" as determined by the layout?

3-12. What are layout machines? How do they facilitate the layout work?

CHAPTER 4

Drilling Machines, Twist Drills, and Auxiliary Tooling

Drilling machines, or drill presses, as they are also called, are used to open, enlarge, and to finish holes by drilling, reaming, boring, counterboring, countersinking, and tapping. These operations are also performed on other machines such as lathes, boring mills, milling machines, and precision jig borers. Many different types of drilling machines have been built for handling different classes of work to the best advantage. Some designs incorporate great versatility in handling a wide variety of different workpieces, while others facilitate the mass production of duplicate parts. Many modern drilling machines are numerically controlled; others are automatically cycled. The horsepower of drilling machines may range from a fractional horsepower for drilling small holes to as much as fifty horsepower for driving a heavy-duty, multiple-spindle drill head that is capable of drilling many holes simultaneously.

Upright Drilling Machine

Upright drill presses, illustrated in Figs. 4-1 and 4-2, are general purpose drilling machines; yet, they can be readily adapted to do production work. The upright drill press shown in Fig. 4-1 has a round column and circular table; however, some upright drilling machines, such as the one shown in Fig. 4-2, have a box column and a square table. The circular table can be raised and lowered. It can also be rotated about the column and about its own center. This feature provides a very convenient method of aligning workpieces that are clamped to the table, with respect to the spindle. The workpieces are clamped with bolts anchored in T slots that are machined in the table. An American Standard Morse Taper is used to hold and drive twist drills, drill chucks, drill sleeves, reamers, and other cutting tools. The spindle housing may be raised or lowered and it contains the feed mechanism. An automatic power feed mechanism is provided in order to feed the twist drill through the work. Different spindle speeds are obtained by shifting the control lever that is on the head of the machine.

Sensitive Drilling Machine

The machine in Fig. 4-2 is a sensitive upright drilling machine. This machine is used for drilling smaller holes and it is not provided with a

Courtesy of The Fosdick Machine Tool Company

Fig. 4-1. Upright drilling machine—round column, circular-table type.

Courtesy of The Fosdick Machine Tool Company
Fig. 4-2. Sensitive upright drilling machine—box column, square-table type.

Courtesy of the Clausing Division of The Atlas Press Company

Fig. 4-3. Bench type, sensitive drill machine.

power feeding mechanism. Instead, it is designed to provide a fine sense of feel to the spindle feed lever which aids the operator in feeding the drill through the hole. It is similar in construction to an upright drill press although its features are lighter. A bench-type, sensitive drill press is shown in Fig. 4-3. Very small, sensitive drill presses capable of drilling extremely small holes are built to do instrument work.

Radial Drilling Machine

Radial drilling machines, Fig. 4-4, are very useful for drilling holes in larger workpieces. In this illustration, holes are being drilled in a large machine frame which is resting on the floor plate, or base, of the radial drill press. The arm of the radial drilling machine is supported by a large, round column. The arm can be raised and lowered, and it can be rotated around the axis of the column. The spindle head, or drilling head, is mounted on the arm. It contains the spindle, the spindle speed-change mechanism, and the automatic power feed mechanism. The spindle head can be moved lengthwise along the arm and clamped in any desired location. Thus, the spindle moves in an arc around the center of the column, and radially, in or out from the column. This design affords great flexibility and ease in positioning the drill in the exact location desired on the workpiece. The arm of the radial drill press provides a means of

Ch. 4 DRILLING MACHINES, TWIST DRILLS 65

Courtesy of the Giddings & Lewis Machine Tool Company

Fig. 4-4. Radial drilling machine drilling large machine frame which is resting on base of drill.

extending the spindle over a workpiece in order to drill holes in surfaces that cannot be reached by the ordinary drill press, as shown in Fig. 4-4.

Turret Drilling Machines

Some drilling machines are equipped with a turret having six faces. Each face has a spindle which can hold a drill or other hole-producing

cutting tool. The turret is indexed manually or automatically to bring each spindle into the operating position with respect to the work. Thus, a number of operations can be performed on a hole, or a group of holes, without the need to change cutting tools as would be the case if a single-spindle drilling machine was used.

Multiple Spindle and Gang Drilling Machines

These machines are used for high production work. Multiple spindle machines have two or more spindles built on a single head. All of the drills feed into the work together, resulting in the production of two or more holes, simultaneously. As many as thirty or more spindles have been

Courtesy of the Clausing Division of The Atlas Press Company

Fig. 4-5. Gang drilling-machine equipped with air-hydraulic spindle feeds. Multiple drill-heads are mounted on two of the spindles.

built into a single head. Special multiple spindle heads have been designed which can be attached to the spindles of upright drilling machines. Multiple-spindle drilling machines are also called "way-type" drilling machines; the name being derived from "ways" on which the multiple drill heads slide. One to four ways can be built into one machine, each way containing a multiple drill head. The ways can be designed to cause the drill heads to slide at any angle from vertical to horizontal.

Gang drilling machines consist of two or more drilling machine spindles that are positioned vertically over a single table. Each spindle contains a different cutting tool and the work is moved from one spindle to the next. Gang drill presses are, in effect, upright drilling machines with two or more spindles. Figure 4-5 illustrates a gang drilling machine. Multiple spindle heads are mounted on two of the spindles and the work is held in a box jig which is fastened to the table. The work is moved from one spindle to another to complete all of the drilling operations.

Twist Drills

The basic metal-cutting tool for opening holes in metal is the twist drill. Its apparent simplicity and the ease with which it can be used tend to obscure the fact that it is a complicated cutting tool with a very complex cutting action. The twist drill must withstand high torque and thrust forces, as well as the temperatures resulting from the formation of the chip. The various elements of a twist drill are defined below and in Fig. 4-6:

Axis—The imaginary straight line forming the longitudinal center of the drill.

Body—That portion of the drill extending from the shank, or neck, to the outer corners of the cutting edge.

Flutes—The helical grooves cut, or formed, in the body of the drill to provide cutting lips, to permit removal of chips, and to allow cutting fluid to reach the cutting lips.

Land—The peripheral portion of the body between adjacent flutes.

Body Diameter Clearance—That portion of the land that has been cut away so that it will not rub against the walls of the hole.

Margin—The cylindrical portion of the land which is not cut away to provide clearance.

Web—The central portion of the body that joins the lands. The extreme ends of the web form the chisel edge. The thickness of the web is not uniform, but increases from the point toward the shank.

Point—The cutting end of the drill, made up of the ends of the lands and the web. In form it resembles a cone, but departs from a true cone to furnish relief behind the cutting lips in order that they can penetrate into the metal and form a chip.

Point Angle—The angle included between the cutting lips, projected upon a plane parallel to the drill axis, and parallel also to the two cutting lips.

Fig. 4-6. Nomenclature of a twist drill.

Lips—The two cutting edges extending from the chisel edge to the periphery.

Relief—The result of the removal of tool material behind or adjacent to the cutting lip and leading edge of the land, to provide clearance; prevent heel drag; and to allow the cutting lips to penetrate into the work and form the chip.

Lip Relief Angle—The axial relief angle at the outer corner of the lip. It is measured by projection into a plane, tangent to the periphery, at the outer corner of the lip.

Chisel Edge—The edge at the end of the webb that connects the cutting lips.

Chisel Edge Angle—The angle included between the chisel edge and the cutting lip, as viewed from the end of the drill.

Helix Angle—The angle made by the leading edge of the land with a plane containing the axis of the drill. This angle is also the rake angle of the drill.

Shank—That part of the drill by which it is held and driven.

Straight Shank—A shank having the form of a cylinder. Usually provided on drills that are to be held in a chuck.

Taper Shank—A shank having an American Standard (Morse) Self-Holding Taper.

Tang—The flattened end of a taper shank, intended to fit into a driving slot in a socket or a machine spindle.

Tang Drive—Two opposite, parallel driving flats on the extreme end of a straight shank.

Table 4-1. Suggested Lip Relief Angle

Drill Sizes	Drill Diameter Range, inch	Lip Relief Angle
#80 to #61	(.0135 to .0390)	24°
#60 to #41	(.0400 to .0960)	21°
#40 to #31	(.0980 to .1200)	18°
1/8 to 1/4	(.1250 to .2500)	16°
F to 11/32	(.2570 to .3438)	14°
S to 1/2	(.3480 to .5000)	12°
33/64 to 3/4	(.5156 to .7500)	10°
49/64 and larger		8°

Courtesy of The Metal Cutting Tool Institute

Suggested lip relief angles are given in Table 4-1. The chisel edge angle is also of great importance. Failure to provide a chisel edge angle that is large enough, will reduce or completely eliminate the relief behind the cutting lips, in the area near the chisel edge. The chisel edge angle has a direct relationship to the lip relief angle and should be 125° to 130° for lip reliefs of from 8° to 10°; 135° for lip reliefs of from 12° to 15°; and 130° to 145° for small drills under 3/32 inch. The point angle is sometimes modified when drilling different materials. Usually, harder materials re-

Table 4-2. Recommended Drill-Point Angles

Work Material	Recommended Point Angle (Included Angle)
General purpose, plain carbon steels; annealed alloy steels; gray cast-iron; medium-hardness, pearlitic, malleable iron.	118°
Stainless steel, alloy steel, heat-treated steel, drop forgings, hard-to-cut materials.	125°
Soft cast-iron; ferritic, malleable iron, aluminum; plastics; fiber; marble; slate; wood; hard rubber.	90° to 110°
Soft and medium copper, brass.	100° to 118°
Magnesium alloys. (Grind flat on face of cutting lips.)	60° to 118°
Soft aluminum, die castings, plastics, fiber, wood.	60°
7 per cent to 13 per cent manganese steel, tough alloy steels, armor plate, extra hard materials. (Grind flat on face of cutting lips.)	150°

quire larger point angles and softer materials are cut faster and more efficiently with a more acute point angle. Recommendations for point angles when drilling various materials are given in Table 4-2. The point angle must be symmetrical with respect to the axis of the drill and both lips should be the same length; otherwise, the drilled hole will be oversized and the tendency to drill a crooked hole will increase. Drill points can be ground manually; however, for the best results, a drill point grinding machine should be used. A theoretically perfect drill point must have the lip relief angle increased as the center of the drill is approached. The Oliver drill pointer, shown in Fig. 4-7, will grind a theoretically perfect point. This machine can also grind a point having a small nib at the center of the chisel edge which causes the drill to be self centering when starting. This feature is especially useful when drilling on numerically controlled machine tools.

The Cutting Action Defined

The cutting action of the chisel edge requires that a relatively large thrust force be used to cause the drill to penetrate into the workpiece. The thrust force can produce severe stresses and strains in the drilling machine, the workpiece, and in the twist drill itself. This action is especially severe on drills that have been shortened by repeated grinding at the point, since the web of twist drills is made thicker toward the shank. The thrust force required to cause the drill to penetrate can be significantly reduced by thinning the web of the drill at the point. The best way to do this is to use a machine such as the drill point thinner (shown in Fig. 4-8) which accurately and uniformly grinds inside both flutes in order to thin the web. Some of the types of web thinning that can be ground

Courtesy of the Oliver Instrument Co., Inc.

Fig. 4-7. Oliver drill point grinding machine. This grinder will grind the lip relief angle increasing as the center of the drill is approached.

on this machine are shown in Fig. 4-9. The ideal way to thin the point is shown at 1 in Fig. 4-9.

This method thins the web, maintaining a uniform web thickness, and it corrects the cutting edges for index by making them identical in length and angle on both sides of the axis of the drill. At 2, the grind extends into the cutting edge, producing a double cutting edge. A chipbreaker type thinning is shown at 3. A small, half-round groove produces a steep hook along the full length of the cutting edge, which tends to curl and to break up the chip. At 4, the web is thinned by grinding a flat along the entire length of the cutting edge. The flat, in effect, reduces the

Courtesy of the Oliver Instrument Co., Inc.
Fig. 4-8. Oliver drill point thinning machine.

rake angle of the drill, strengthens the cutting lips, and reduces or eliminates the tendency of the drill to seize when breaking through the bottom side of a through hole. The point shown at 5, is called the split point or the crankshaft point. It is also shown in Fig. 4-10. This point was originally developed for drilling oil holes in automotive crankshafts; thus the name "crankshaft-type point." The split point reduces the thickness of the web to a minimum. Two additional cutting edges are ground at the point; they provide a nice cutting action and an easier escape route for the metal being removed in this region. The drill is ground in the conventional manner after which the angle is ground to 55° with respect to a plane that is perpendicular to the axis of the drill. For the point shown at 6 in Fig. 4-9, the web has been thinned by grinding away some of the relief behind the lips. The cutting lips are not disturbed with this grind; and, for this reason, this method is recommended when the web must be thinned by hand grinding. When needed, the advantages of thinning the web are so great that this should be done by hand if a machine is not available to do the job.

Ch. 4 DRILLING MACHINES, TWIST DRILLS 73

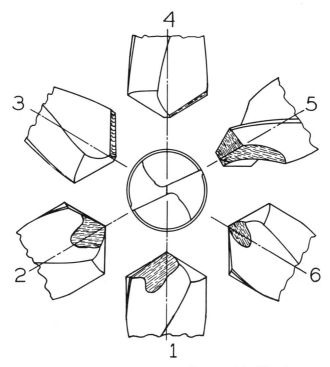

Courtesy of the Oliver Instrument Co., Inc.

Fig. 4-9. Points ground by drill point thinning machine.

Courtesy of the National Twist Drill Company

Fig. 4-10. Split-point drill grind. The thickness of the web is reduced to a minimum by this grind.

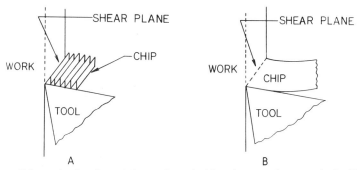

Fig. 4-11. Schematic drawing of formation of chip when cutting metal. A. Discontinuous chip. B. Continuous chip.

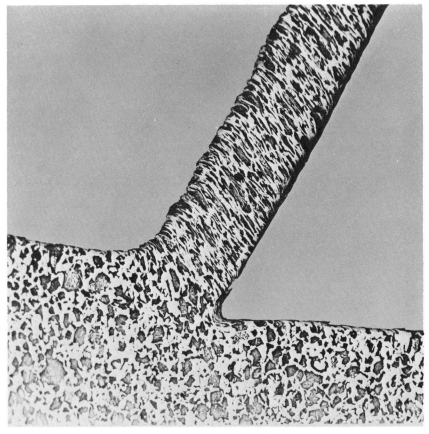

Courtesy of the National Twist Drill Company

Fig. 4-12. Photomicrograph of a metal chip formed by a cutting edge such as the lips of a twist drill.

Chip Formation in Drilling

When metals are subjected to the action of a cutting edge that is harder and stronger than the metal that is being cut, a chip is formed. As the cutting edge engages the workpiece, the material ahead of the cutting edge is compressed and escapes by flowing in a direction parallel to the face of the tool. As the cutting edge continues its penetration, the force on the tool and on the work increases until it is great enough to cause either rupture or shear within the grains of the metal, on a plane that is roughly perpendicular to the face of the tool. This plane is called the shear plane and is shown in Fig. 4-11. The shear plane is usually a somewhat curved surface rather than a perfectly flat surface. The chips may be discontinuous as at A in Fig. 4-11, or they may be continuous as shown at B. Brittle materials, such as cast iron, tend to form discontinuous chips while ductile materials, such as soft steel, tend to form continuous chips. Figure 4-12 is a photomicrograph of a ductile steel, showing the distortion of the grains along the shear plane in the forming of a continuous chip. The lips of the twist drill form a chip in exactly this manner.

The total cutting action of a twist drill, however, is more complex due to the action of the chisel edge, shown in Fig. 4-13. At the exact center of the chisel edge the only motion of the drill is such that the action is similar to a center punch that penetrates the metal by extruding it sideways. As the distance from the center increases, the rotation of the drill becomes effective and the chisel-edge wedge appears both to cut and extrude the

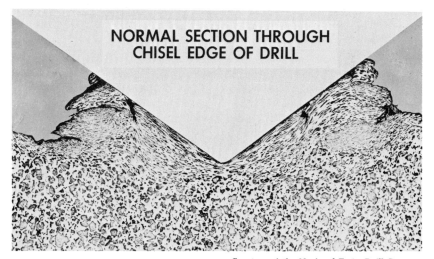

Courtesy of the National Twist Drill Company

Fig. 4-13. Photomicrograph through the axis of a drilled hole normal to the drill chisel-edge showing the complex cutting and extruding action by which the metal in this region is removed.

DRILLING MACHINES, TWIST DRILLS — Ch. 4

Table 4-3. Thrust in Pounds for Drilling AISI 1112 Steel

Drill Size, Inches	Feed per Revolution, Inch								
	.001	.002	.004	.006	.008	.010	.013	.016	.020
	Thrust, in pounds								
⅛	23	42	75	108					
¼	45	82	145	215	265	325			
⅜		120	215	305	390	475			
½		150	270	385	490	600	740	890	1100
⅝			320	450	585	700	880	1050	1300
¾			375	530	660	820	1025	1200	1500
⅞			430	610	760	950	1175	1400	1750
1				700	890	1100	1350	1650	2000
1¼				880	1100	1400	1750	2100	2550
1½				1100	1400	1750	2200	2600	3200

Courtesy of The Metal Cutting Tool Institute

metal. The very complex metal deformation can be seen by the deformation of the grains under the chisel edge in Fig. 4-13. This severe deformation causes the high thrust force required by twist drills. The magnitude of these thrust forces is given in Table 4-3. As previously mentioned, the thrust force can be considerably reduced by thinning the web or by grinding a split point on the drill.

There are then two distinct chip formation actions in drilling. This can be readily seen in Fig. 4-14, which clearly shows the chip formed by

Courtesy of the National Twist Drill Company

Fig. 4-14. Section of hole with drill removed to show separate chip formed by the lip and the chisel edge.

Ch. 4 DRILLING MACHINES, TWIST DRILLS

the lip, as distinct from the chip formed by the chisel edge. In this respect, twist drills are unique among the metal-cutting tools.

Drill Sizes

Twist drill sizes are given in fractional inches, number sizes, and letter sizes. The fractional inch sizes start at $1/64$ inch (.0156) and continue to a diameter of $3\frac{1}{2}$ inches. Number sizes begin with No. 97 (.0059 inch) and continue to No. 1 (.228 inch). Letter size drills start with A (.234 inch) and end with Z (.413 inch). Under average shop conditions, a twist drill will drill holes that are somewhat larger than its diameter. The amount that a drill will cut oversize is given in Table 4-4. When very

Table 4-4. Oversize Amount Normally Cut by a Drill Under Normal Shop Conditions, in Inches

Drill Dia, Inch	Amount Oversize, Inch		
	Average Max	Mean	Average Min
$1/16$.0020	.0015	.0007
$1/8$.0045	.0027	.0010
$1/4$.0066	.0041	.0025
$1/2$.0080	.0048	.0029
$3/4$.0082	.0052	.0032
1	.0092	.0067	.004

Courtesy of The Metal Cutting Tool Institute

accurate hole diameters are required, other machining methods are employed; although, generally, the first step in machining an accurate hole is to drill the hole.

Reamers

Figure 4-15 illustrates a chucking reamer and a hand reamer, and it defines their respective individual elements. Reamers are used to finish cut holes to a precise dimension and to provide a smooth surface finish on the walls of the hole. They are precision cutting tools that must always be handled with care to prevent chipping and other damage to the reamer teeth. *One cardinal principle in using a reamer is never to rotate it in the reverse direction.* Reamers should always be rotated in the forward, or cutting direction, at all times; regardless of whether they are entering or leaving the hole. Failure to do this will rapidly damage the margin and destroy the accuracy of the reamer. Reaming is a fast, efficient, and easily applied method of finishing holes.

There are two basic types of reamers: machine reamers and hand reamers. Machine reamers may have a straight, or a taper, shank. They are used primarily in power-driven machine tools. Machine reamers have a chamfer ground on the ends of the lands. The chamfer angle is usually 45°, although other chamfer angles are sometimes used. The lands have a

Fig. 4-15. Reamer nomenclature.

Courtesy of the Metal Cutting Tool Institute

margin adjacent to the cutting edge, which is unrelieved. Relief is ground on the back of the chamfer, thereby making it into a sharp cutting edge. The chamfer edges do the cutting and remove the excess metal from the hole. The forces on the workpiece resulting from the cutting action of the reamer will cause the work to spring slightly away from the reamer, in the region of the chamfer edge. After the chamfer edge has passed a given portion of the hole, the work material will spring back and bear against the margin of the lands of the reamer. The margin, being unrelieved, is a poor cutting edge. Primarily, it burnishes the sides of the hole to produce a good surface finish. It may also scrape a very small amount of

metal out of the hole, thereby removing small irregularities. The margin also performs another important function in supporting and steadying the reamer as it passes through the hole. Sometimes, the margin is ground with a slight taper toward the chamfer edge. The length of this taper is from ⅛ to 3⁄16 inch and the angle, with respect to the axis of the reamer, is from 1° to 5°. This is sometimes called a lead. Actually, its function is to increase the burnishing action of the lands, thereby improving the surface finish of the hole.

Hand reamers are designed to be turned manually in reaming a hole. They are ground with a starting taper on the lands which is relieved to form a sharp cutting edge. (See Fig. 4-15.) A bevel edge is also ground on the end, however most of the cutting is done by the starting taper. Hand reamers are designed to remove less metal from a hole than machine reamers will. The margin of the hand reamer also burnishes the hole in the same manner as does the margin on the machine reamer.

The speed and feed through the hole of a hand reamer are controlled entirely by the workman. On machine reamers, the speed and feed are controlled by the machine setting. As a general rule, the cutting speed for reaming should be about one-half to two-thirds of the speed used for drilling the same material. Excessive cutting speeds will cause the reamer to chatter; consequently, the cutting speed must be kept low enough to prevent the occurrence of chatter. The feed for machine reaming is usually .0015 to .0040 inch per flute per revolution. Thus, the feed rate of a six-fluted reamer that is cutting at 100 rpm, using .004 inch per tooth feed, would be 6 × .004, or .024 inch per revolution. Expressed in different terms, the feed rate would be 100 × .024, or 2.4 inches per minute. When the feed used is too low the work may be glazed, the cutting edge will wear excessively, and, occasionally, chatter will occur. An excessively fast feed rate will reduce the accuracy of the hole and lower the quality of the surface finish.

It is very difficult to generalize on the amount of stock that should be removed by reaming as this is dependent upon the nature of the work material, the design of the reamer, the feed used, the finish required, and the depth of the hole. As a starting point for machine reaming, allow .008 to .012 inch on a ¼-inch hole, .012 to .016 inch on a ½-inch hole, and .025 to .035 inch on holes up to 1½-inches in diameter. For hand reaming, the stock allowance is much smaller. The allowance ranges from .001 to .010 inch.

Two other types of reamers that are sometimes used are rose reamers and taper reamers. Rose reamers are used in a machine. They are similar to machine reamers in construction; however, they are designed to remove much more metal from a hole. They do not ream a hole as accurately or with as good a surface finish as do the standard machine reamers. Taper reamers are used to produce taper holes. Larger size reamers are often made as shell reamers. The shank, or arbor, of the shell reamer is separate from the body of the reamer. The advantage of this design is that reamers of different size can be held on the same

arbor and when a body is worn out the arbor does not require replacement.

Counterbores, Countersinks, and Spotfacers

Counterbores, countersinks, and spotfacers are frequently used in drill-press work in order to modify an existing hole. The counterbore, A, in Fig. 4-16, is used to enlarge the end of a hole by cutting a cylindrical surface that is concentric with the original hole. The concentricity is obtained by the pilot located on the end of the counterbore and which acts to guide the tool as it penetrates into the work. A countersink, B, in Fig. 4-16, is used to make a cone-shaped enlargement in the end of the hole. A spotfacing tool is shown at C. The purpose of the spotfacing operation

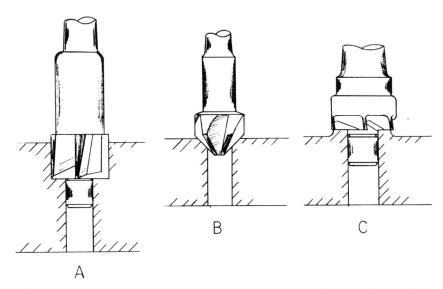

Fig. 4-16. A. Counterboring a hole. B. Countersinking a hole. C. Spot-facing a hole.

is to cut a smooth, flat surface which is perpendicular to the axis of the hole. Usually, this surface serves as a seat for the head of a cap screw or bolt; or as a seat for a nut.

Taps

The basic nomenclature of a tap is shown in Fig. 4-17. A tap is used to produce internal threads. It has thread-shaped teeth which cut a thread of similar form when it is screwed into a drilled hole. The first step in tapping an internal thread is to drill a hole with a tap drill. This is an ordinary drill that is appropriately sized so that it will leave just the right amount of metal in the hole for the tap to remove and finish the thread forms to required size. The tap is then screwed into the work-

Ch. 4 DRILLING MACHINES, TWIST DRILLS 81

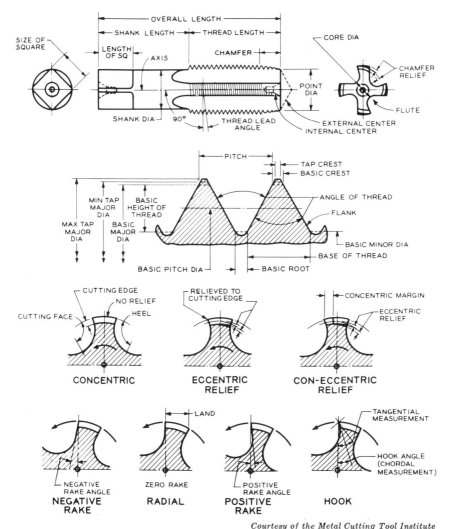

Fig. 4-17. Basic nomenclature of a tap.

piece by hand, with the aid of a tap wrench, or by power, as applied by a machine tool such as a lathe or a drilling machine. The feed of the tap must always be equal to the lead of the thread being cut. The lead of a thread is the distance that it advances in one revolution. A tap will generally feed itself, once it has started, as it will tend to follow that portion of the thread that already has been cut in starting.

The size of the tap drill to be used to tap the hole is dependent upon the diameter of the thread, the size of the thread tooth (or pitch of the thread), the material of which the thread is made, the length of engage-

ment of the thread, and the available standard drill sizes. For ordinary manufacturing, the internal thread in a tapped hole need not be more than 75 or 80 per cent of the theoretical, standard thread depth. For some classes of work, not more than 50 per cent of the thread depth is required. The diameter of the tap drill should not be smaller than is necessary to give the required strength of thread, as every decrease in the diameter of the tap drill increases the power required for tapping, and, incidentally, the percentage of broken taps. Recommended tap drill sizes and hole sizes for tapping Unified Screw Threads can be found in various standard machine shop handbooks such as Machinery's Handbook. An old rule of thumb used where a table of tap drill sizes is not available is:

To find the tap drill diameter, in inches, subtract from the outside diameter an amount equal to 1 divided by the number of threads per inch.

Example 4-1:

Find the tap drill diameter for a $\frac{3}{4}$-10 unified thread. This thread has 10 threads per inch; hence,

$$\text{Tap Drill Diameter} = \tfrac{3}{4} - \tfrac{1}{10} = .65 \text{ inch}$$

In the practical application of this rule, the nearest commercial drill size is always used. In this particular case, a $2\tfrac{1}{32}$-inch drill is almost the exact equivalent of .65 inch.

The taps used for tapping operations are usually made in sets of three, and are known as taper, plug, and bottoming taps or Nos. 1, 2, and 3. These taps are distinguished by the amount of chamfer on the end. The angle formed between the chamfer (see Fig. 4-17) and the axis of the tap is called the "chamfer angle." This angle determines the number of threads which have their height reduced on the end of the tap.

Taper taps have a chamfer angle of 4° which reduces the height of approximately 8 to 10 threads; plug taps have a 9° chamfer angle with 3 to 5 threads reduced in height; and bottoming taps have a 30° chamfer angle with 1½ threads reduced in height. The chamfered teeth are relieved behind the face of the thread tooth and all of the cutting is done by them. The number of flutes and the length of the chamfer determine the chip load, or the amount of stock, that must be removed by each tooth. Holes which extend through the workpiece can be tapped with either a taper or a plug tap. When hard or tough materials are encountered it is sometimes advisable to use a taper and a plug tap alternately; thus, when a taper tap sticks after being screwed into a hole it is removed, and the plug tap inserted, to relieve the long cut made by the taper end of the first tap. Bottoming taps are primarily used to finish the bottom of blind holes, or holes closed at the bottom, after a taper or plug tap has cut as deeply into the hole as possible. Not all blind holes are finished with a bottoming tap. Often, the tap drill hole can be drilled somewhat deeper than the depth required to be threaded and the hole can be finished with a plug tap.

A cutting fluid should always be used with all materials when tapping except on cast iron and plastics. When hand tapping a hole, the tap should be advanced approximately one turn and then rotated in the reverse direction until the chip is broken. By repeating this sequence until the hole is tapped, a smooth thread will be produced with minimum effort. Machine-driven taps are generally screwed into the hole while rotating in the forward direction and are reversed to remove the tap when the job is finished. On very difficult jobs the tap is sometimes reversed when cutting into the hole in a manner similar to hand tapping.

Spade Drills

Spade drills, Fig. 4-18, are frequently used to drill large diameter holes. They are made in sizes ranging from one to five inches. A spade drill

Courtesy of the Metal Cutting Tool Institute

Fig. 4-18. Elements of a spade drill.

consists of a holder made from heat-treated steel and a high-speed steel, or cemented-carbide, tipped blade. The blade is seated in a slot in the end of the holder which is clamped together by a screw to secure the blade in place. Each holder can hold a range of different blade sizes. The shank of the holder may be straight or have a standard Morse taper. Some spade drill holders have coolant holes drilled lengthwise which supply a stream of coolant close to the cutting edges. Notches or chip breakers are ground on the cutting edges to reduce the width of the chip. These notches are unevenly spaced so that they do not occur at the same radial distances on the two cutting edges. This is necessary to prolong the life of the cutting edges. A chip curler ground behind the cutting edge on the

face of the blade provides a rake angle and tends to curl and break up continuous chips as they are formed.

The spade drill may be used by rotating the drill as in a drill press, or by feeding it through a rotating workpiece, such as when drilling on a lathe. The same cutting speed and feed are normally used with a spade drill as are used with a twist drill. Cast iron can be drilled to almost any depth with a spade drill. When drilling steel or aluminum, problems with heat and chip removal can occur if the hole depth is more than one and one-half, or two times the diameter, of the drill. This can be remedied by providing an adequate flow of coolant into the hole.

Some of the advantages of spade drills are listed below:

1. Low initial cost and low cost of replacement blades.
2. Intermediate size blades can easily be made up in order to produce nonstandard size holes.
3. Flat end blades are available for producing flat-bottomed holes or for forming internal shoulders.
4. Special corner radii and chamfers can easily be formed in a hole by grinding the blade to suit.
5. The spade drill has a relatively small chisel edge which reduces the magnitude of the thrust force required to push the drill through the hole.
6. Because of the low thrust force it is possible to drill large holes in one pass which otherwise would require drilling progressively larger-sized holes until the required size is obtained.
7. A short spade-drill holder can be used which reduces the twisting action or torsional deflection of the drill.
8. Long spade-drill holders can be used to drill deep holes which are beyond the reach of ordinary twist drills.
9. When a short spade-drill holder is used, its resistance to bending is increased which improves the tendency to drill a straight hole.

Gun Drills

Gun drills are customarily used to drill deep holes where the length of the hole is five or six times its diameter. These are special drills and are usually used on special deep-hole drilling machines. They have a single flute and a single cutting edge which may be high-speed steel or a brazed-on cemented carbide tip. The cutting edge may be designed to cut to the center of the drill. Some gun drills have a cutting edge that does not cut to the center of the drill, thereby leaving a core of metal which can be easily removed if the hole is a through hole. This method of producing, where metal is left in the center, is called *trepanning*. In either case the angles on the cutting edge are designed to make the drill produce a straight hole. The body of the gun drill is made from crimped tubing which allows a cutting fluid to be fed through the drill to an oil hole near the cutting edge. Gun drills must be started with the aid of a bushing.

Once they are well started, however, they are self-supporting and will drill a straight hole. When using a gun drill in a lathe the need to use a guide bushing to start the drill, can be eliminated by first drilling and reaming a hole about one diameter deep in order to start the gun drill.

When a deep hole is required and a gun drill is not available, the hole can be drilled by brazing an extension onto the shank of a standard twist drill. The hole is first drilled about one diameter deep with a standard-length twist drill, and the remainder of the hole is drilled with the drill having the extended shank. This is a slow process because the drill must be backed out of the hole frequently, in order to remove the chips.

Thrust and Power for Drilling

The thrust required for drilling AISI 1112 steel is given in Table 4-4. This table lists the thrusts measured by a dynamometer for different drill sizes operated at different feed rates. A conversion factor, given in Table 4-5, allows the data in Table 4-4 to be applied to calculate the thrust required to drill other materials.

Example 4-2:

Find the thrust required to drill AISI 3150 steel with a ⅝-inch drill using a feed of .013 inch per revolution.

1. The thrust required to drill AISI 1112 steel with a ⅝ drill, using a .013 inch per revolution feed, is 880 lbs, as obtained from Table 4-3.
2. The conversion factor for AISI 3150 steel, as obtained from Table 4-5, is 1.4.
3. Calculate the thrust.

$$\text{Thrust for AISI } 3150 = 880 \times 1.4 = 1{,}232 \text{ lbs}$$

Table 4-5. Approximate Conversion Factors for Thrust to Drill Other Materials (Use with Table 4-4.)

Materials	Thrust Conversion Factors	
	Regular Points	Split Points
MST 6A1-4VA Titanium Alloy BR.340	2.0	0.8
17-7 PH Stainless Steel BR.400	2.3	1.0
4340 Heat Treated Steel 240,000 to 260,000 psi	3.5	
AISI 1020 Steel	1.4	
AISI 1035 Steel	1.3	
1.00 per cent Carbon Tool Steel	1.7	
AISI 3150 Steel	1.4	
Malleable Iron	0.6	
Gray Cast-Iron	0.6	

Courtesy of The Metal Cutting Tool Institute

Table 4-6. Horsepower at the Cutter (hp$_c$.) Required to Drill AISI 1112 Steel at 100 RPM

Drill Size, Inches	Feed Per Revolution, Inch								
	.001	.002	.004	.006	.008	.010	.013	.016	.020
	Horsepower at Cutter								
⅛	.002	.004	.006	.009					
¼	.007	.013	.023	.030	.037	.042			
⅜	.016	.027	.048	.068	.080	.094			
½	.027	.042	.081	.115	.135	.163	.20	.25	.28
¾	.06	.10	.18	.25	.32	.38	.45	.52	.62
1	.10	.17	.31	.42	.52	.64	.76	.90	1.1
1¼	.15	.26	.45	.64	.78	.94	1.15	1.35	1.6
1½	.22	.36	.62	.88	1.1	1.3	1.6	1.9	2.2

Courtesy of The Metal Cutting Tool Institute

Drill-point wear, or improper sharpening of the drill point, can increase the thrust by 100 per cent or more. When the values obtained from Tables 4-4 and 4-5 are used, proper allowance should be made to anticipate the maximum conditions.

The horsepower at the cutter, or drill, required to drill AISI 1112 steel using a cutting speed of 100 rpm, and at various conditions, is given in Table 4-6. It must be emphasized that these values are for the actual power used at the drill and that it does not consider the losses in power transmission from the motor to the spindle. This loss is usually accounted

Table 4-7. Approximate Conversion Factors, *F*, for Horsepower at the Cutter (hp$_c$.) Required to Drill Other Materials (Use with Table 4-6.)

Material	Conversion Factor, *F*
AISI 1020 Steel	1.6
AISI 1035 Steel	1.3
AISI 3150 Steel	1.6
1.00 per cent Carbon Tool Steel	1.7
AISI 4340 Steel Heat-Treated to 240,000 to 260,000 psi	2.3
Gray Cast-Iron	.5
Malleable Iron	.6
AISI 416 Free-Machining, Martensitic Stainless Steel	1.2
AISI 303 Free-Machining, Austenitic Stainless Steel	1.6
AISI 304 Austenitic Stainless Steel	1.8
17-7 PH Precipitation-Hardened, Austenitic Stainless Steel	2.0

Courtesy of The Metal Cutting Tool Institute

Ch. 4 DRILLING MACHINES, TWIST DRILLS

for by applying a machine efficiency factor, that for drilling machines, averages from 70 per cent for a geared head to as high as 85 to 90 per cent for a direct belt drive. Table 4-7 lists conversion factors for calculating the horsepower required to drill other materials. To calculate the horsepower use the following formulas:

$$\text{hp}_c = \text{hp}_{c'} \times F \times \frac{N}{100} \qquad (4\text{-}1)$$

where: hp_c = Horsepower at the cutter for the material to be drilled.
$\text{hp}_{c'}$ = Horsepower at the cutter for AISI 1112 steel as obtained from Table 4-6.
F = Conversion factor for material to be drilled from Table 4-7. (Note: $F = 1$ when AISI 1112 is drilled at cutting speeds other than 100 rpm)
N = Cutting speed in revolutions per minute (rpm) to be used in drilling.

$$\text{hp}_m = \frac{\text{hp}_c}{e_m} \qquad (4\text{-}2)$$

where: hp_m = Horsepower at the motor of the drilling machine
e_m = The mechanical efficiency of the machine.

Example 4-3:

Calculate the horsepower at the cutter (hp_c) and at the motor (hp_m) required to drill a ½ inch hole in AISI 416 stainless steel at 75 rpm with a .006 inch per revolution feed rate.

1. The horsepower at the cutter required to drill AISI 1112 steel with a ½-inch drill at .006 ipr is .115 as obtained from Table 4-5.
2. The conversion factor, F, as obtained from Table 4-7, is 1.2.
3. Calculate the horsepower at the cutter.

$$\text{hp}_c = \text{hp}_{c'} \times F \times \frac{N}{100}$$

$$= .115 \times 1.2 \times \frac{75}{100} = .103$$

4. Calculate the horsepower at the drilling-machine motor. Assume a machine efficiency of 80 per cent.

$$\text{hp}_m = \frac{\text{hp}_c}{e_m} = \frac{.103}{.8} = .13$$

Again, this does not provide an allowance for drill-point wear or for improper sharpening. When making equipment estimates it is recommended that these values be doubled.

QUESTIONS

4-1. What is the advantage of the round column-circular table type of upright drilling machine?

4-2. What is the advantage of the radial drilling machine?

4-3. Describe the difference between a turret drilling machine, a multiple-spindle drilling machine, and a gang drilling machine.

4-4. Define the following elements of a twist drill:

body	relief	point angle
flutes	shank	lip relief angle
land	webb	chisel edge angle
margin	point	tang

4-5. Why is it necessary to grind the correct chisel edge angle on the twist drill?

4-6. When is it advisable to grind a flat on the face of the lips?

4-7. Name the characteristics and the advantages of a machine ground, as compared to a hand-ground drill point.

4-8. Describe how a continuous chip is formed when cutting metal.

4-9. Describe how a discontinuous chip is formed when cutting metal.

4-10. How does the chisel edge of a drill form a chip?

4-11. What is the advantage of thinning the webb of a twist drill?

4-12. How does the cutting action of a hand reamer differ from that of a machine reamer?

4-13. What function does the margin of a reamer perform?

4-14. What is the cardinal principle in using any reamer?

4-15. A machine reamer having six flutes is cutting at a speed of 125 rpm. The feed should be exactly .002 inch per flute per revolution.

 a. What should the feed of the machine be if it is set in terms of inch per revolution?

 b. What should the feed of the machine be if it is set in terms of inch per minute?

 c. If the hole is $1\frac{1}{2}$-inches deep, how long will it take the reamer to pass through the hole on the cutting stroke?

4-16. How much stock should be left in a hole for:

 a. A machine reamer?

 b. A hand reamer?

4-17. Calculate the size of the tap drill required for a $\frac{1}{2}$-13 threads per inch tap. Check this calculation with the tap drill size recommended in the appendix.

4-18. Describe the effect of the chamfer angle on the performance of a tap.

Ch. 4 DRILLING MACHINES, TWIST DRILLS 89

4-19. Calculate the thrust required to drill a one-inch hole in 1% Carbon Tool Steel using a feed of .016 inch per revolution.

4-20. The drill press used to drill the tool steel part in Problem 4-19 has a 2½-horsepower motor. If the cutting speed is 125 rpm, and the machine efficiency of the drill press is 80 per cent, can this job be done on this machine? Prove by calculating the horsepower required at the motor.

CHAPTER **5**

Drilling Machine Operation

Work on machine tools is accomplished in the following sequence:
1. Plan how to do the job.
2. Obtain all the tools and materials required to do the job.
3. Prepare the workpiece and the machine for the machining operation.
4. Start the job, proceed with thought and care, and get the job done.

Although the basic principles of machine tool operation are unvarying, the manner in which they are applied can vary from shop to shop. The reason for this is that the resources (men, tools, machine tools, and experience) of each individual shop are different. The skilled craftsman adapts his knowledge and skills to the resources that are available in the shop where he is employed. It is, therefore, not unusual for the same job to be done differently in various shops; however, regardless of the method used, the basic principles must be applied to each method. In this book the basic principles are emphasized. The student must, however, be prepared to adapt them to the circumstances with which he is confronted.

The object of all drilling machine work is to machine holes to a given size having an acceptable surface finish and located exactly where required on the workpiece. Drilling machines, or drill presses as they are also called, are primarily hole-producing machines unlike lathes or milling machines which machine a variety of shapes including holes.

A great many drilling jobs are preceded by a layout operation on the workpiece. Layout work was covered in Chapter 3 and will not be discussed in this chapter. Where a layout is required, it is assumed that this work has been done prior to the discussion in this chapter.

Drilling Speeds

For any machining operation, including drilling, the cutting speed is generally given in terms of feet per minute, which is abbreviated: fpm. It describes the rate of movement of a cutting tool relative to the workpiece. For example, a cutting speed of 100 fpm would describe a cutting tool traveling in a straight line at a uniform rate of motion, or speed, such that the tool would travel a distance of 100 feet in one minute. The effect would be the same if the tool was stationary and the workpiece

Ch. 5 **DRILLING MACHINE OPERATION** **91**

moved past the tool at this speed. In drilling, the tool moves, but it rotates instead of moving in a straight line. The surface on the lips adjacent to the margin is the part of the drill traveling at the highest speed when the drill is rotating. This surface travels a distance equal to the circumference of the drill, or $\pi \times$ drill diameter, every time that it makes one complete turn. Assume for a moment that the drill diameter is ½ inch, its circumference would be ½ × 3.1416 = 1.57 inches. Now, if this drill can cut at a cutting speed of 100 fpm, or 12 × 100 = 1200 inches per minute, the number of circumferences that will fit into this length is $\frac{1{,}200}{1.57}$ or 764 circumferences per minute. Since every circumference also represents one complete revolution of the drill, the ½-inch drill must turn 764 revolutions per minute if the outer edge of the lip is to move a distance of 1,200 inches per minute, or 100 feet per minute. To summarize this as a formula:

$$\text{rpm} = \frac{12\,V}{\pi D} \quad (5\text{-}1)$$

where: rpm = Revolutions per minute at which the drill must operate. This is the speed setting to be made on the drill press.
 V = Cutting speed in feet per minute (fpm), at which the drill can cut the work material.
 π = pi = 3.1416. The mathematical symbol for this number which is a constant for calculating the circumference of a circle.
 D = Diameter of the drill, in inches.

Some drilling machines have infinitely variable speeds within a range of speeds while others have discrete speeds which are in steps, such as are available in a geared head or on a step cone pulley. When a machine is equipped with a stepped range of speeds, it should be set at the speed that is the nearest to that calculated in Formula 5-1.

Example 5-1:

Two holes are to be drilled in a part made from AISI 4140 steel that has been quenched and tempered to have a hardness of 275 Bhn (Brinell hardness number). The diameter of one hole is ⅛ inch and the diameter of the second hole is 1 inch. The cutting speed for AISI 4140, heat treated to the hardness given, is 50 fpm when the drill is made from high-speed steel.

A. Calculate the cutting speeds for both drills.

 1. For the ⅛-inch drill:

$$\text{rpm} = \frac{12\,V}{\pi D} = \frac{12 \times 50}{3.14 \times \tfrac{1}{8}} = 1{,}528$$

2. For the 1-inch drill:

$$\text{rpm} = \frac{12 \times 50}{3.14 \times 1} = 191$$

B. The drill press has the following speeds (rpm): 1,200, 800, 600, 500, 400, 200, 150, 90. What speeds should be selected to drill the holes in part A?

1. For the $\frac{1}{8}$-inch drill—1,200 rpm
2. For the 1-inch drill—200 rpm

Although some carbon-tool-steel drills are still made for home use, most drills used by industry are made from high-speed steel. High-speed steel can cut about two times as fast as carbon tool steel. The cutting speed that can be used in drilling is dependent upon the material from which the twist drill is made, the material from which the workpiece is made, the hardness of the work material, and the heat treatment of the work material. The cutting speed for reaming should be from one-half to two-thirds of the cutting speed for drilling. A table of cutting speeds for drilling and reaming is given in the appendix.

Drilling machines are usually designed so that the feed of the spindle is arranged in increments that are expressed in terms of inches per revolution, abbreviated: ipr. The feed rate for drilling is dependent upon the nature of the work material (type, hardness, and heat treatment) and upon the size of the twist drill. Recommended feeds for drilling are given in the appendix. All of the information necessary to select the cutting speed and feed for drilling, is not always available. Often, the hardness and the heat treatment of the metal being drilled is not known; in which case judgment must be used, and the tables approached as a starting point from which changes are made as experience dictates.

Drilling Setups

The basic conditions for drilling a hole may vary considerably as determined by circumstances, due to the machine, the workpiece, and the judgment of the machinist. Sometimes, the workpiece is clamped firmly in a fixed position and is immovable, as shown at A, in Fig. 5-1. At other times, the work is not clamped to the table but is held to the table by the operator's hand, and is free to move and center itself with respect to the drill, as shown at B and C, Fig. 5-1. Another condition was shown in Fig. 4-4. The heavy machine base being drilled is immovable due to its own weight; however, the operator can unclamp the arm and the head of the radial drilling machine so that the drill can move freely in order to align itself for drilling. It is usually desirable to allow some movement, or "float," between the twist drill and the workpiece in order that they may align themselves. An exception to this is machine tools that are capable of accurately aligning the spindle that drives the drill to the position on the work where the hole is required; such as in the case of jig borers

Ch. 5 DRILLING MACHINE OPERATION 93

Courtesy of the Clausing Division of the Atlas Press Company

Fig. 5-1. (Upper left) Part clamped directly to the table. (Upper right) Bolt clamped to table used to resist torque when drilling a larger part. (Lower left) Small part held by parallel clamps. (Lower right) Drilling ornamental iron part.

and numerically controlled drilling machines. Most ordinary drill presses depend upon the operator to obtain the required alignment.

It is not possible to tell in drilling, exactly when a part must be clamped rigidly to the table of the machine, or when it can be held to the table by hand. It is always best to err in the direction of safety when this judg-

ment has to be made. In making this decision the following must be considered:

1. Is the twist drill small enough to generate a relatively low torque when drilling?
2. Is the part to be drilled large enough so that there is enough leverage to overcome the torque manually?

A very serious accident can be caused by a workpiece being caught and spun around by a drill that is rotating at a high speed, or even, at times, at lesser speeds.

Figure 5-2 illustrates a part that is being held in a precision drill-press vise, for drilling. Note the stepped jaws to provide clearance at the bottom to allow the drill to penetrate through the hole without damaging the vise. The vise has a quick locking mechanism for locking the movable jaw. Attached to the drill press is also a "Slide Lox" mechanism which allows the vise to move on the drill press table. When the drill is centered with the workpiece, the vise is quickly locked into position by this

Courtesy of Lassy Tools Inc.

Fig. 5-2. Part is being drilled while held in vise. Vise is mounted on "Rapid Lox" base enabling it to be quickly locked in position for drilling.

Ch. 5 DRILLING MACHINE OPERATION 95

Courtesy of the Clausing Division of the Atlas Press Company

Fig. 5-3. A. Work clamped to angle plate for drilling. B. Work being drilled at an angle. Note use of strap clamps.

mechanism. Larger and heavier workpieces are sometimes allowed to "float" on the drill press table in order to align the spindle to the center punch mark made by the layout as shown at B, Fig. 5-1. To resist the drill torque, however, a bolt is clamped to the table through one of the

slots against which the workpiece is positioned. When drilling a hole clear through the workpiece, or a "through" hole, care must be exercised not to drill into the drill press table. This can be avoided by positioning the table with respect to the spindle so that the twist drill is over an opening in the table. Otherwise, parallels must be placed between the table and the part, as at A, Fig. 5-4. The magnitude of the torque is usually not too great when drilling a small hole; however, it is still unsafe to hold the part to the table by hand if the part is small. The part should be held in a small vise which, in turn, can be held to the table by hand. Sometimes the part can be held in parallel clamps instead of a small vise, as at C, Fig. 5-1. Parts that are not conveniently held in a vise, and cannot be held to the table by hand, must be clamped directly to the table as shown at A, Fig. 5-1. Ingenuity often must be used in making a setup as shown at D, Fig. 5-1. The ornamental iron part is placed over the piece of wood which is clamped to the table with a C-clamp. The end of the wood has been roughly shaped on the end to the radius of the ornamental iron part. At A, Fig. 5-3, the part is clamped to an angle plate that is bolted to the drill press table. The setup for drilling a hole at an angle with respect to the base of the part is illustrated at B, Fig. 5-3. The drill press table is tilted to the required angle and the part is held to the table with strap clamps.

Strap clamps are frequently used to clamp workpieces to machine tool tables. Although they are relatively simple, some understanding is necessary to use them correctly. A correct-size heel block should be used so that the strap clamps are parallel to the machine table when they are positioned against the work. A step block, shown at B, Fig. 5-3, sometime is convenient in obtaining the height required to keep the strap clamp level. A second important principle in their use is to always position the bolt as close to the workpiece as conditions will permit. The proportion of the clamping force exerted by the bolt that is applied to the workpiece is in direct relation to the distance between the bolt and the workpiece. One of the clamps shown in the illustration is a gooseneck clamp. The other clamp has a finger, or pin, on the end which is used to clamp inside the hole. Generally, clamps should not be used on the inside of finished holes unless that hole is protected by soft metal shims. Sometimes holes are drilled into castings for the sole purpose of providing a seat for the fingers of strap clamps.

The spindle of the radial drill press can be positioned in areas on the workpiece which are sometimes difficult, or impossible, to reach with an upright drill press. Such a setup is shown at A, Fig. 5-4, which illustrates the punch plate of a die block being drilled. In this setup the punch plate is placed on parallels that are positioned on the drill press table. The parallels will provide clearance for the drill when it penetrates the punch plate and prevent it from damaging the table top. It is not necessary to clamp the punch plate to the table as its weight is sufficient to resist the drill torque and prevent it from moving as the hole is being drilled. Smaller workpieces, as in B, Fig. 5-4, must be clamped to the table to

Ch. 5 DRILLING MACHINE OPERATION

Courtesy of The Fosdick Machine Tool Company

Fig. 5-4. Typical radial drilling machine setups. (Upper left) Weight of large die setpunch holder, sufficient to resist drill torque. Work is placed on parallels, to avoid drilling into table. (Upper right) Work clamped to table top. (Lower left) Cylindrical part clamped to V-notch on side of table. (Lower right) Table set at angle for drilling.

prevent their movement when the holes are drilled. The hole being drilled is not a through hole; thus, it is not necessary to use parallels as at A. Note the use of strap clamps in this setup. At C, Fig. 5-4, a round bar of steel is bolted against a V-notch (not visible) machined into the side of the table. The V-notch aligns the bar in a vertical position for drilling holes in the end, as shown. The table of the radial drill press may be set

at an angle to drill an angular hole as shown at D, Fig. 5-4. This part is clamped with "homemade" strap clamps which are thick steel straps through which a hole has been drilled.

A job that is frequently encountered is that of drilling a hole through, and perpendicular to, the axis of a cylindrical part, as shown in Fig. 5-5.

Lower left view courtesy of the Clausing Division of the Atlas Press Company

Fig. 5-5. (Upper left) Using the blade of a combination square to align cylindrical part. (Upper right) Using the beam of a combination square to align a cylindrical part. (Lower left) Setup for clamping cylindrical part in V-block. (Lower right) Using a center head combined with a square head to align a large cylindrical part.

The workpiece is first laid out as described in Chapter 3. (See E, Fig. 3-4.) The part is placed in a V-block, or, if it is a long part, in a matched pair of V-blocks. It is positioned so that the drill will not cut into the V-block. Some V-blocks have a hole through the center for drill clearance. When two V-blocks are used the drill can be positioned between them. The scribed layout line on the end face of the workpiece is aligned in a vertical position by the blade, or the beam, of a combination square,

Fig. 5-6. Method of starting drill, concentric with scribed circle.

as shown at A and B, Fig. 5-5. The work is then clamped into position, as shown at C. The square is not used to align the drill with the workpiece. Very large cylindrical parts can sometimes be aligned in the manner shown at D. The blade of a combination set has a center head and a square head fastened to it, as shown. The blade is positioned over the punch mark at the center of the hole. Sometimes, greater accuracy can be obtained by placing the blade opposite the layout line as shown at D. The workpiece is adjusted until the spirit level on the square head is exactly level. The part is then clamped into position.

Drilling the Hole

The operation of drilling the hole can start when all of the preparations have been completed; i.e., the work is set up, a sharp drill of the correct size is held in the spindle of the machine, and the machine has been set up for the correct cutting speed and feed. The first step is to align the drill with respect to the center punch mark at the center of the hole to be drilled. The drill point is then fed into the workpiece until it has penetrated approximately two-thirds of the drill's diameter. The drill is then backed away and the relation of the drilled spot to the scribed layout circle should be observed. At A, Fig. 5-6, the spot produced by the drill point is not concentric with the layout circle, indicating that if the hole were to continue to be drilled in this position it would not be in the desired location. To correct this situation, cut one or more shallow grooves on the heavy side, as shown at B, Fig. 5-6, with a gouge, or draw chisel (Fig. 5-7). The groove unbalances the cutting forces acting against the lip of the drill causing it to shift in the direction of least resistance, which is toward the grooved side. The drill point is fed a little deeper into the work and the concentricity between the spot and the layout circle is again checked. If necessary, additional grooves are cut with a draw chisel. When the spot and the layout circle are concentric, as

Fig. 5-7. Draw chisel or gouge.

shown at C, Fig. 5-6, the hole can be drilled to the required depth. When the hole has been drilled, about one-half of each witness mark will remain to show that the hole is drilled exactly where the layout has indicated that it be made. It should be mentioned that it is usually not necessary to cut grooves in every hole as described. Frequently, the initial check of concentricity between the spot and the layout circle will show that they are concentric and the hole can be drilled to depth without additional positioning of the work.

An alternate method of starting a drilled hole, preferred by some, is to center drill a hole first (see Fig. 5-8). The center-drilled hole is then used as a guide for starting the twist drill. The center drill is short, stiff, and will not bend. Furthermore, it has a relatively small chisel edge when compared to larger size drills, thereby enabling the center drill to pene-

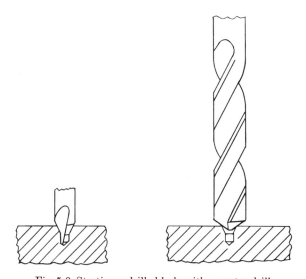

Fig. 5-8. Starting a drilled hole with a center drill.

trate the workpiece with greater ease. The coned surface formed by the center drill has an included angle of 60° which is steeper than the included angle of the drill point. Thus, the lips of the twist drill contact the workpiece first and the chisel edge does no work until the location of the hole is established. The point angle of the lips will cause the drill to follow the axis of the hole made by the center drill, and, since the chisel edge of the twist drill does not engage the work, the drill will penetrate into the work with ease. This method is especially recommended for accurate work on numerically controlled machines, jig borers, milling machines, and other machine tools where the machine spindle is accurately located with respect to the location of the hole on the workpiece. However, on conventional drill presses, when the workpiece is too heavy

to "float" on the table and to align the center punch mark to the center drill, this method will not give accurate results. In this instance the use of the previous method is recommended because any misalignment in starting the drill can be corrected with the draw chisel. The disadvantage of center drilling is that it is very difficult to correct any misalignment once the hole has been drilled with a center drill. Another disadvantage of this method is that it requires a frequent tool change.

Table 4-3 shows that the thrust required to feed a larger size drill is very high. A large proportion of this thrust is due to the action of the chisel edge. The thrust required can be greatly reduced and the drill made to cut faster, by first drilling a smaller "lead hole" as shown in Fig. 5-9.

Fig. 5-9. A small lead hole is sometimes used to reduce the thrust required for a large drill.

Reaming

Reamers are used to finish drilled holes to size and to a smooth, surface finish. In order to produce accurate holes by reaming, the reamer must be sharp and correctly ground, the right amount of stock must be left in the hole for reaming, and the reamer must be accurately aligned with the axis of the drilled hole. The feed of the reamer through the hole must not be too fast (.0015 to .004 inch per land), otherwise the accuracy and the surface finish of the hole will be impaired. Sometimes a reamer will chatter. This chatter may exist only at the start of the hole as the reamer aligns itself with the drilled hole. Chatter is objectionable because it impairs the surface finish of the hole and it can damage the surface finish of the reamer. Cemented-carbide tipped reamers, in particular, are quite likely to chip if chatter occurs even at the start of the hole.

To eliminate chatter the following steps should be tried: reduce the cutting speed; vary the feed; reduce the relief angle on the reamer; use

a piloted reamer; use a reamer having helical flutes; increase the rigidity of the setup; and, as a last resort, use a reamer having irregular spacing of the flutes. The use of a helical fluted reamer is much preferred to one having irregular spacing. The helix angle of helical fluted reamers should not be greater than necessary, as this will tend to increase the force required to feed the reamer through the hole. One exception to this are taper reamers; these will cut better with a very steep spiral. Often, the difficulties associated with chatter can be eliminated by exercising great care in aligning the reamer with the hole, at the start of the operation. This should never be done carelessly.

A good supply of the correct cutting fluid should be used for the reaming operation except when reaming cast iron, which should be reamed dry, with only a jet of compressed air to act as a cooling medium. Use of the correct cutting fluids often materially improves the surface finish obtained in a reamed hole. Generally, for reaming steel, a sulfurized mineral oil will work best although good results have also been obtained using soluble oil. Soluble oil and lard oil compounds can be used on aluminum. Brass may be reamed dry, although the application of a soluble oil or a lard base, oil compound is often advisable. Soluble oil, or a sulfurized mineral oil, is recommended for stainless steels. Many proprietary cutting fluids can be purchased which will do an excellent job.

Tapping

When tapping is to be done in a drilling machine there must be provision for rapidly reversing the spindle so that the tap can be backed out of the hole after the thread is cut to the required depth. Most radial drilling machines are designed with a spindle reverse mechanism that is actuated by the spindle control lever. Upright and bench drilling machines are usually not equipped with such a mechanism for reversing the rotation of the spindle thus a special attachment for this operation is required. An automatic reverse tapping attachment is shown in Fig. 5-10. This attachment has a taper shank which is inserted in the spindle of the machine. The body of the chuck and the gage B, which may be used to control the depth to which holes are tapped, are both prevented from rotating with the machine spindle by rod A. When tapping a hole the spindle is lowered by hand and the tap is fed to the required depth manually. When the lower end of the stop rod B, comes into contact with the face of the work, the forward rotation of the tap is stopped. When the drill press spindle is raised, gears within the chuck body are engaged that cause the tap to back out of the hole rapidly. The tapping attachment may be equipped to have an adjustable friction drive that stops the rotation of the tap whenever the tapping torque exceeds a preset limit, thus safeguarding the taps against breakage.

There are many more variables that must be considered when selecting the cutting speed for tapping than for other machining operations; thus, it is not surprising that the best and most efficient cutting speeds for tapping cannot be tabulated with any assurance that the tap will then

Fig. 5-10. Automatic tapping attachment applied to upright drilling machine.

be operating at its highest efficiency. Among the factors that must be considered are:

1. Material to be tapped
2. Heat treatment and hardness of material to be tapped
3. Length of hole
4. Length of chamfer on top
5. Pitch of thread
6. Percentage of full thread to be cut
7. Cutting fluid to be used, and
8. Design of the drilling machine with respect to the sensitivity of the controls.

For example, speeds must be lowered as the length of the hole is increased because long holes accumulate chips, increase friction, and interfere with the flow of cutting fluids. Taps having long chamfers can be run faster than taps with short chamfers in short holes; however, in long holes the taps with the short chamfers can be run faster. Bottoming taps must be run slower than plug taps. A slower speed is required to tap full-depth thread than a 75-per cent-depth thread. When the tap diameter exceeds approximately ½ inch, coarse thread taps should be run more slowly than fine thread taps having the same diameter. Taper threaded taps, such as pipe taps, should be operated from one-half to three-quarters of the speed of a straight tap of comparable diameter. Automatic machines in which the movements required for tapping are machine controlled, can, in most cases, be operated at greater tapping speeds than manually operated machines. The machine operator must exercise good judgment in deciding the limits of his ability to control the machine in order to prevent tap breakage and other possible damage. A table of

recommended tapping speeds is given in the appendix, but these may have to be modified by the factors discussed above.

When tapping, a cutting fluid should always be used with all materials except cast iron and plastics. Cast iron and plastics are machine tapped dry or with a jet of compressed air to cool the tap and help to remove the accumulation of chips. It is not practical to give specific recommendations for tapping other materials because of the wide variety of proprietary cutting fluids that cán do an excellent job. The following recom-

Courtesy of the Clausing Division of the Atlas Press Company

Fig. 5-11. Casting held in a drill jig for drilling operation. Note that the part is held to the table manually.

Ch. 5 DRILLING MACHINE OPERATION 105

mendations are general and it should be understood that there are exceptions. Sulfur based oils with active sulfur are recommended for tapping plain carbon steels, alloy steels, malleable iron, and Monel metal. Tool steels, high speed steels, stainless steels, and alloy steels that have been heat treated to a higher hardness should be tapped using a chlorinated sulfur base oil. Mineral oil with a lard base is recommended for aluminum, brass, copper, manganese bronze, naval brass, phosphor bronze, and Tobin bronze. Aluminum and zinc die castings should be tapped with lard oil that has been diluted with 40 to 50 per cent kerosene. The proper cutting fluid used when tapping can increase tap life, improve surface finish, reduce friction, and result in better size control.

Drill Jigs and Fixtures

Drill jigs and fixtures are frequently used when a number of duplicate parts have to be drilled. The shop terms are understood to have the following meaning:

> Drill fixtures are specially designed tools that are used to hold the workpiece in a fixed location. They may incorporate a method or locating device to establish a known relationship between locating, or reference, surfaces on the part and the spindle of the machine. Drill fixtures have come into prominence with the advent of numerically controlled drilling machines.
>
> Drill jigs incorporate all of the features of drill fixtures and, in addition, they are constructed with guide bushings that guide the twist drill or the reamer in the required relationship with respect to the workpiece. Drill jigs are used on all types of drilling machines except numerically controlled machines.

Drill jigs provide for a fast and accurate method of locating the drill in the position on the workpiece where the hole is wanted, without requiring a layout to be made, and without the need to use the draw chisel or a center drill to start the hole. There are many different types of drill jigs, some of which are very simple, while others are quite complex. The basic functions of a drill jig are to locate the workpiece in the position required for drilling, to clamp it in this position, and to guide the drill

Fig. 5-12. Drill jig for drilling bushing.

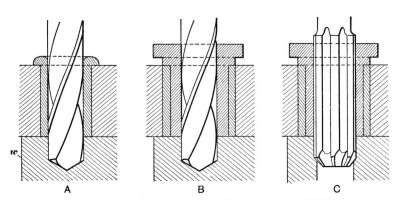

Fig. 5-13. Fixed and removable guide bushings for drill jigs.

by means of hardened steel bushings. An example of the use of a drill jig is shown in Fig. 5-11. The part is located and clamped in the drill jig which is held to the table, manually. The jig can be positioned under the drill quickly and easily, for drilling several holes. Another drill jig is shown in Fig. 5-12. The workpiece is shown at A, and the jig is used to drill the ¼-inch hole. The work is shown in position in the jig by the dashed and dotted line. Figure 5-13 illustrates the application of hardened-steel drill jig bushings. These bushings fit the drill closely and keep it in proper position. Some jigs have fixed guide bushings and others have removable guide bushings. A fixed bushing is shown by the sectional view at A, Fig. 5-13, which also indicates how the drill is guided while it is drilling the work, w. Jigs are equipped with removable bushings should drills of a different size need to be used or if drilled holes are to be finished by reaming. For example, if a hole is to be drilled and reamed, a removable bushing is used that fits the drill, as shown at B, and this is replaced by a bushing that fits the reamer, as shown at C.

Drill fixtures are similar to the drill jigs shown except that they do not have drill bushings. For this reason they are generally more open in construction than drill jigs. They are always fastened to the table of the machine so that their relationship to the spindle remains fixed. Drill jigs, on the other hand, may or may not be fastened to the table.

QUESTIONS

5-1. Why might an identical job be done with equal efficiency by different methods in two different machine shops?

5-2. Calculate the cutting speed in rpm at which a drilling machine must be set in order to drill the following size holes in AISI 1040 steel of 180 Bhn hardness:

$\frac{1}{16}, \frac{1}{4}, \frac{1}{2}, \frac{3}{4}, 1, 1\frac{1}{2}$

Ch. 5　　　DRILLING MACHINE OPERATION　　　107

5-3. A drill has the following speeds given in rpm: 260, 530, 1,060, 1,860, 3,600. At what speed should it be set in order to drill the following size holes in AISI 4140 steel which is heat treated to have a hardness of 230 Bhn:

$$\tfrac{1}{16}, \tfrac{1}{4}, \tfrac{1}{2}, \tfrac{3}{4}, 1, 1\tfrac{1}{2}$$

5-4. A drill press is to be designed to drill 17-7 PH Precipitation Hardened Austenitic Stainless Steel. The range of drill sizes for which the machine is to be designed is to be from $\tfrac{1}{8}$ to 1 inch. The minimum hardness is to be 150 Bhn and the maximum hardness is to be 400 Bhn. What should be the minimum and the maximum rpm of this drill press?

5-5. From what material are twist drills made?

5-6. What factors determine the cutting speed (fpm) for drilling?

5-7. Define what is meant by inches per revolution (ipr) feed rate.

5-8. Upon what is the feed rate for drilling dependent?

5-9. Why is it often desirable to allow some movement or "float," between the twist drill and the workpiece, when drilling?

5-10. What must be considered when deciding whether or not to clamp the workpiece to the table?

5-11. What must be done when setting up a drill press for drilling a through hole?

5-12. Name two important principles to be observed when applying strap clamps to clamp a part to the machine-tool table.

5-13. What are V-blocks and for what purpose are they used?

5-14. What is the draw chisel and how is it used to correct misalignment between the drill and the hole?

5-15. How is the center drill used in starting a hole?

5-16. When should a center drill not be used to start a hole?

5-17. Name at least seven steps that should be taken to help eliminate chatter when reaming a hole in a drill press.

5-18. What are the variables that must be considered in selecting the cutting speed at which a tap should operate when tapping in a drill press?

5-19. What is the difference between a drill jig and a drill fixture?

5-20. What are the advantages of using drill jigs in drilling and reaming holes?

CHAPTER **6**

Engine Lathe Construction

The origin of the term "engine lathe" undoubtedly can be traced to the manner in which early metal-cutting lathes were powered. These early lathes were driven by a reciprocating steam engine through a system of pulleys, belts, and clutches. Usually, a single steam engine provided the power for several lathes as well as for the other machine tools in the machine shop. Later, a single electric motor replaced the steam engine

Courtesy of The R. K. LeBlond Machine Tool Company

Fig. 6-1. Large lathes at work.

Ch. 6 ENGINE LATHE CONSTRUCTION 109

Courtesy of the Atlas Press Company

Fig. 6-2. Modern 12″ bench lathe.

to drive a group of machine tools. All modern engine lathes, as well as other machine tools, are driven by an individual electric motor. The individual motor drive eliminates the need for overhead shafting, reduces power losses, provides a more direct drive, and provides greater flexibility in the placement of the lathe on the machine shop floor.

Engine lathes differ from other lathes by their greater simplicity of construction and by their greater versatility in performing many different machining operations. Turning cylindrical surfaces, facing flat surfaces, cutting threads, drilling and boring holes, and cutting internal threads are some of the typical operations that are performed on engine lathes. Of equal importance in considering the versatility of the engine lathe is the ease with which the workpiece can be set up and held, and the ease with which the lathe can be manipulated in performing different operations. Although production lathes and numerically controlled lathes are used, to a large extent, where a variety of parts must be made, the engine lathe is still the basic machine tool in the machine shop. It is used extensively in jobbing shops and in tool and die shops. When operated with skill and with an understanding of its capabilities, it can produce parts to exacting standards of accuracy and finish. When the appropriate size of machine is used, the engine lathe can produce both large and small parts. A group of large engine lathes at work is shown in Fig. 6-1. As a contrast in size, a small bench lathe is shown in Fig. 6-2. A modern 10″ precision lathe is shown in Fig. 6-3. The construction and principles of operation of both large and small lathes is essentially the same.

Principal Features

The principal parts of an engine lathe are shown in Fig. 6-4. The bed of the lathe, Fig. 6-5, is a heavy and rigid casting made from gray cast-iron. Gray cast-iron is used as a material of construction for machine

110　ENGINE LATHE CONSTRUCTION　Ch. 6

Courtesy of the Clausing Division of the Atlas Press Company

Fig. 6-3. Modern 10″ lathe.

Courtesy of The R. K. LeBlond Machine Tool Company

Fig. 6-4. The principal parts of an engine lathe.

Courtesy of The Monarch Machine Tool Company
Fig. 6-5. Lathe bed showing the ways of the lathe.

tool beds and frames because it has excellent vibration dampening qualities and it is a good bearing material for sliding bearings. The bed rests on two legs with the exception of small bench lathes which have the bed attached directly to the bench. The top surfaces of the bed are accurately machined to provide a sliding surface for the carriage and to provide for the precise alignment of the headstock and the tailstock. This alignment is achieved by providing "ways" which are raised surfaces machined on the top of the bed. The ways are machined parallel to each other. One of the two ways is used to align the headstock and the tailstock. The other way is used to guide the movement of the carriage as it slides over the bed causing it to be parallel to the axis of the lathe. The direction of this carriage movement is called longitudinal. Some lathes, that shown in Fig. 6-2, for example, have flat beds that do not have ways. The surfaces on the sides of the bed that are perpendicular and adjacent to the top of the bed are carefully machined parallel to each other. These surfaces perform the function of the ways on flat bed lathes.

Methods of Drive

The lathe is driven by an electric motor that is usually located inside the cabinet leg at the headstock end. A multiple V-belt drive transmits the power from the motor to the headstock. The principal function of the headstock is to support and to align the headstock spindle. On many lathes it also contains a selective gear transmission that provides for different speeds of the lathe spindle. A typical geared headstock is shown

Courtesy of The Monarch Machine Tool Company

Fig. 6-6. Geared type headstock.

in Fig. 6-6. The spindle speeds on geared head lathes are changed by placing the spindle speed change levers, located on the side of the headstock, in the position prescribed by the instruction plate located behind these levers. (See Fig. 6-7.) The speed change mechanism on many smaller lathes consists of stepped pulleys of different diameters as shown in Fig. 6-8. The belt can be on any of the four different pairs of pulleys that

Courtesy of The Monarch Machine Tool Company

Fig. 6-7. Geared type headstock and feed-thread gearbox.

Ch. 6 ENGINE LATHE CONSTRUCTION 113

provide four different speeds to the lathe spindle. Four additional spindle speeds are obtained through a two-step pulley arrangement on the drive from the electric motor to the countershaft, which contains one of the stepped pulleys, or sheaves, that drives the spindle. Eight additional spindle speeds are obtained by engaging the back gears, thus providing

Courtesy of the Atlas Press Company
Fig. 6-8. Stepped pulley type of headstock.

the lathe with a total of 16 spindle speeds. The back gears consist of four gears located inside the headstock that reduce the speed of the stepped pulley in the headstock. When the back gears are engaged this pulley is disconnected from the headstock spindle; when the back gears are disengaged this pulley is connected to the headstock and drives it directly. Another type of headstock construction is shown in Fig. 6-9. A spindle-speed mechanism, located outside the headstock (not shown in Fig. 6-9), may be a stepped pulley arrangement, a gear transmission, or a variable-speed drive. The back-gear drive can be seen in this illustration. It provides the slow spindle speeds.

114 ENGINE LATHE CONSTRUCTION Ch. 6

Courtesy of the Clausing Division of the Atlas Press Company

Fig. 6-9. Modern headstock showing back gears and antifriction spindle bearings.

The Headstock Spindle

The lathe headstock spindle is mounted on machine-tool grade anti-friction bearings which can be seen in Fig. 6-9. These bearings are manufactured to closer dimensional tolerances than ordinary anti-friction bearings so that the runout, or eccentricity, of the spindle as it rotates is held to a minimum. The spindle is hollow to allow bar stock and lathe attachments to pass completely through, if desired. The spindle nose is very accurately machined to provide locating surfaces for lathe centers, faceplates, and chucks. The inside of the spindle nose (Fig. 6-10) has a taper which is ground to a high degree of accuracy and surface finish. This taper provides a seat for an adapter which is used to reduce the size of the taper required by the headstock center. The headstock center is held by the taper in the adapter. The construction of the inside of the spindle nose for all engine lathes is the same, except for size. There are three different designs for the outside of the spindle noses, and these are shown in Fig. 6-11. At the left is the long, tapered type of spindle nose. The spindle nose mountings (chucks and faceplates) are located by the taper and are driven by the key. The threaded ring screws onto a matching thread on the chuck and faceplate, and it holds them securely on the taper. The center view (Fig. 6-11) shows a cam-lock type of spindle nose. It has a short, tapered surface which locates the spindle mountings. The chucks and faceplates have a cam stud (see Fig. 6-10) which fits into each of the holes located above the taper. The cam is turned by a chuck key, to hold the spindle mounting firmly, on the short taper. The

Ch. 6 ENGINE LATHE CONSTRUCTION 115

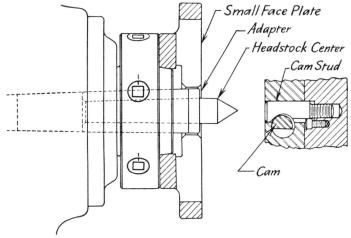

Fig. 6-10. Cam-lock type spindle nose showing adapter and headstock center in place.

cam studs also act as drivers for the mounting. A threaded spindle nose is shown in the view to the right, in Fig. 6-11. The spindle nose mounting is held in place and is driven by the threads. The shoulder behind the threads is used to help locate the mounting on the spindle nose. The spindle nose should be kept free of nicks and scratches, and it should be carefully cleaned before a chuck or a face is mounted on it. Also, the

Fig. 6-11. (Left) Long taper type spindle nose. (Center) Cam-lock type spindle nose. (Right) Threaded spindle nose.

corresponding surfaces of the chuck or faceplate must be clean and kept free from any damage.

The Feed-Thread Gearbox

The feed-thread gearbox, also called the quick-change gearbox, is located on the side of the lathe below the spindle (see Fig. 6-7). It is connected to the spindle by a gear train which can be seen in Figs. 6-4, 6-8, and 6-9. Provisions are usually made to enable some of the gears in this train to be changed in order that threads having special leads, such as metric threads, can be cut. The function of the feed-thread gearbox is to provide the lathe with the capability of having different rates of automatic power feed and to enable it to cut threads of different pitches or numbers of threads per inch. It is, essentially, a selective-speed-ratio transmission box which changes the ratio of the speed of the lathe spindle with respect to the lead screw and feed rod (see Fig. 6-4). The feed rate, or number of threads per inch, for which the lathe is to be set up to cut is obtained by positioning the feed-thread change handles according to instructions given on the feedplate.

The output shafts of the feed-thread gearbox are the feed rod and the lead screw. The function of the feed rod is to actuate the automatic-power feed mechanism located in the lathe apron. The lead screw is used to feed the carriage in the longitudinal (parallel to the lathe axis) direction when cutting threads. It should be used only to cut threads in order not to impair its accuracy as the result of wear and the imposition of heavy feeding forces. Some lathes (Fig. 6-4) have a control rod located below the lead screw and feed rod which is used to start and stop the headstock spindle. This rod is actuated by the speed control handle.

The Carriage

The carriage, Fig. 6-12, consists of the saddle, cross slide, compound rest, and apron. The saddle, Fig. 6-13, is an H-shaped casting that slides on the top of the lathe bed. It enables the cutting tool to be moved parallel or perpendicular to the axis of the lathe. The saddle rides on one of the ways on the lathe bed which guides the longitudinal movement on the carriage parallel to the lathe axis.

The Cross Slide

The cross slide moves over the dovetail slide on top of the saddle. The movement of the cross slide is perpendicular to the lathe axis. This movement is actuated by a feed screw, shown in Fig. 6-14. The feed screw can be turned manually by means of the cross-feed handle. The automatic cross feed is actuated by a gear in the apron.

Mounted behind the cross-feed handle is a micrometer cross-feed dial that has accurate graduations cut on its circumference which enables the lathe operator to determine the distance that the cross slide is moved. On many lathes each graduation corresponds to a movement of .0005 inch of the carriage which causes .001 inch to be removed from the diam-

Courtesy of The R. K. LeBlond Machine Tool Company

Fig. 6-12. Principal parts of the lathe carriage.

eter of the workpiece that is being turned, because a cut is taken from both sides of the workpiece as it revolves. Cross-feed dials graduated in this manner are said to be "direct reading" dials. For example, if .050 inch is to be removed from the diameter of the workpiece the cross-feed dial is moved 50 graduations. On some lathes the graduations correspond to a movement of .001 inch, thereby removing .002 inch from the diameter of the workpiece. In this case the amount to be removed from the diameter of the workpiece must be divided by two in order to determine the number of graduations through which the cross-feed dial must be moved. For example, if .050 inch is to be removed from the diameter of the workpiece, the cross-feed dial is moved 25 graduations.

Clearance must be provided between the cross-feed screw and the cross-feed nut (Fig. 6-14) in order to allow the screw to rotate without binding. This clearance is intentional although it results in a certain amount of lost motion between the nut and the screw. Thus, when the cross-feed screw is turned in one direction and then in the opposite direction, it will not move the carriage until the lost motion is taken up. In

Courtesy of The R. K. LeBlond Machine Tool Company
Fig. 6-13. Lathe saddle.

order to accommodate this lost motion, the cutting tool must be positioned by always moving the cross slide toward the workpiece. If the cutting tool must be positioned an exact distance in a direction away from the workpiece axis, the cross slide should be first moved back beyond the desired tool position and then moved forward toward the workpiece until the desired position is reached as determined by the cross-feed dial.

Courtesy of The Monarch Machine Tool Company
Fig. 6-14. Cross-feed screw.

The Compound Rest

The compound rest is clamped to the cross slide and its base can turn through an angle of 360 degrees when the clamp nuts are loosened. The base has graduations enabling it to be clamped at any angle with respect to the lathe axis. The compound-rest slide is moved over the base by a feed screw that is similar in design to the cross-feed screw. The compound-rest feed can only be actuated manually, by turning the compound-rest feed handle. A micrometer dial permits making accurate movements of the compound-rest slide. The compound-rest slide has a large T slot which is used to clamp a tool post or tool block. The cutting tool, or the cutting tool holder, is clamped in the tool post, or tool block. The function of the compound rest is to provide a movement of the cutting tool at any desired angle, with respect to the axis of the work.

The Apron

The apron of the lathe, Fig. 6-15, is fastened to the underside of the saddle. It is essentially a housing that contains the gears and clutches required to actuate the automatic power, longitudinal and croos-feeds. It also supports a large handwheel that actuates the manual longitudinal feed and a lever, shown at the far right, called the "split-nut" lever, which actuates a split nut. The split nut engages the lead screw and drives the carriage when cutting threads. It is disengaged when the thread has been cut to the desired length in order to stop the motion of the carriage. The split-nut lever should never be used to feed the carriage to take any cut other than cutting screw threads. The lever in the center of the apron

Courtesy of The R. K. LeBlond Machine Tool Company
Fig. 6-15. Lathe apron.

is used to selectively engage either the cross feed or the longitudinal feed. The longitudinal feed is actuated by a small pinion at the rear of the apron that engages a rack (see Fig. 6-4) which is located on the side of the bed above the lead screw.

The Tailstock

The tailstock is used to support long workpieces and also to hold drills, reamers, and other cutting tools. A drawing of a tailstock showing all of its parts is illustrated in Fig. 6-16. It consists of a base that rests

Courtesy of the Midwest Industrial Management Association

Fig. 6-16. Details of engine lathe tailstock: 1. Tailstock center; 2. Spindle; 3. Body of tailstock (top); 4. Oil hole; 5. Binder handle; 6. Spindle nut; 7. Screw bushing; 8. Screw; 9. Handwheel; 10. Machine handle; 11. Key for spindle; 12. Clamp bolt; 13. Base; 14. Clamp; 15. Adjusting screws; 16. Alignment scale; 17. Guide; 18. Binder plugs; 19. Binder stud; 20. Woodruff key; and 21. Handwheel nut.

on the bed of the lathe and a body that is clamped to the base. The base, which is aligned with the axis of the lathe by one of the ways of the lathe, has an accurately machined guide that fits the body. The body and the base will slide over the lathe bed until it is in the desired position, then it is clamped to the bed by tightening the clamp bolt nut. The tailstock spindle is prevented from turning, by a key. When the tailstock handwheel is turned, the screw causes the tailstock spindle to slide in or out of the body. The spindle may be clamped by turning the binder handle. The inside of the spindle nose has a taper which is used to hold the tailstock center, drill chucks, or cutting tools such as drills or reamers. This taper must be absolutely clean before a center or other tool is inserted in it. It serves, primarily, to locate; its ability to resist torsion is very limited. It should not be used to hold large drills because the cutting torque of the drill may cause the taper to slip and possibly score the surface. When this occurs, the taper will no longer locate the center or the tool accurately, and it will no longer be able to hold. The repair

of a tailstock spindle is a very expensive procedure. Scored tailstock spindle tapers cannot always be repaired, thus requiring the replacement of the entire spindle. The centers, drill chucks, and cutting tools are removed from the tailstock spindle by backing the spindle into the body until the object that is held in the spindle bumps against the plug end of the screw which forces it out. The body of the tailstock can be moved a limited distance perpendicular to the axis of the lathe by unclamping the clamp bolt nut and turning two adjusting screws located on each side of the base. This movement is used to accurately align the headstock and the tailstock spindles. It can also be used to turn tapers. An alignment scale on the back of the tailstock gives a visual indication of the amount of offset or indicates when the spindle is properly aligned. This scale should not, however, be depended upon for great accuracy.

Engine Lathe Attachments

Many attachments have been developed to increase the usefulness and the accuracy of the engine lathe. Some of these attachments will be discussed in greater detail in subsequent chapters. A group of frequently used engine lathe attachments is shown in Fig. 6-17. A collet chuck and a group of collets are shown at the top of this illustration. A three-jaw universal chuck is shown in the lower left side. Separate jaws for holding

Courtesy of The Monarch Machine Tool Company

Fig. 6-17. Engine lathe attachments.

smaller outside diameters, and for holding on inside diameters, are shown. The jaws of this chuck are actuated by the chuck key. They move in and out together and hold the round workpiece in a central location without the need for further adjustment. A four-jaw independent chuck is shown in the lower right-hand view. Each jaw is adjusted independently of the other jaws by the chuck key which is the T-shaped key adjacent to the chuck. The jaws can be reversed in the chuck body in order to grip onto inside or outside surfaces. A drill chuck is shown between the three- and four-jaw chucks. The drill chuck is used to hold straight-shank twist drills, reamers, and other cutting tools. It is usually held in the tailstock spindle when used in a lathe. A large faceplate is shown in Fig. 6-18. Workpieces are bolted directly to faceplates to perform various turning and facing operations. Lathe fixtures are also bolted to faceplates. A lathe fixture is a special device to hold and to locate workpieces, in a lathe, that cannot be held by other methods.

A micrometer carriage stop is shown in Fig. 6-19. It is used to obtain accurate positions of the carriage and to move it exact distances, longitudinally. A length rod is extended from the headstock. It can be accurately moved in one-inch increments. A micrometer head at the end of the length rod is used to make fine adjustments of less than one inch. The dial test indicator, which is mounted on the carriage, is used to position the carriage so that it will always be at the same distance from the end of the length rod.

A large steady rest is shown in Fig. 6-20. Steady rests are used to support long and slender parts. The steady rest also permits an unobstructed access to the ends of parts so that facing and boring operations can be performed on these ends. The steady rest is composed of a frame containing three jaws. These jaws form a three-point bearing surface and they can be adjusted in or out, radially, by turning the adjusting screws. The frame is hinged on one side allowing the upper half to swing

Courtesy of the Clausing Division of the Atlas Press Company

Fig. 6-18. Large faceplate.

Courtesy of The Monarch Machine Tool Company

Fig. 6-19. A micrometer carriage stop.

open for inserting or removing the work. The base of the frame has V-grooves in it that fit the ways of the lathe bed. When the steady rest is in use, it is clamped to the bed and the jaws are set against the work thus supporting, or steadying it, during the operation.

Tracer attachments can be used to turn and to bore irregular contours, as shown in Fig. 6-21. They can also be used to cut stepped shafts with shoulders and to turn tapers. Internal tapers, internal shoulders, and

Courtesy of The R. K. LeBlond Machine Tool Company

Fig. 6-20. Boring a large workpiece in a lathe with the end held by a steady rest.

Courtesy of The R. K. LeBlond Machine Tool Company

Fig. 6-21. Part turned on engine lathe equipped with tracer attachment.

counterbores can be readily cut with this attachment. The tracer attachment is mounted on the carriage in place of the compound rest, as can be seen in Fig. 6-22. A template with the profile of the part to be turned is mounted on the carriage but is prevented from moving with the carriage. The stylus of the tracer attachment follows the profile of the template which causes the cutting tool to turn this profile on the part in a manner to be described.

The regular cross slide is removed from the lathe and replaced by one having a top slide (similar to the compound rest) mounted on it. This slide is actuated by a hydraulic cylinder which has its movement controlled by a stylus. The nose of the stylus must have the same shape as the nose of the cutting tool. The stylus follows the outline of the template which is mounted on the cross slide and fastened to the bed of the lathe. The position of the template can be adjusted by the micrometer adjustment clamp.

In operation, the first diameter is cut to the required-size by taking a trial cut, using the cross slide to set the tool to the required depth of cut; then the stylus is brought into contact with the template and the automatic power, longitudinal feed is engaged. (The trial cut can usually be eliminated on subsequent parts.) The stylus will then pilot the cutting tool causing it to reproduce the template outline on the workpiece in the following manner: the longitudinal feed moves the carriage toward the tailstock at a steady rate; when the stylus detects a change in the diameter of the part, or a shoulder, the top slide will move in or out at an angle, with respect to the axis of the workpiece. This angular movement of the top slide may be considered as being made up of two component motions: one parallel to the lathe axis and one perpendicular to

Fig. 6-22. Principle of operation of tracer control attachment.

it. For example, when a square shoulder is encountered, the top slide will move at an angle away from the work axis. The rate of movement of that portion of the angular motion of the top slide that is parallel to the lathe axis will be equal and opposite to the longitudinal feed rate. Since these opposite movements are equal and parallel to the lathe axis the tool will, in effect, remain stationary with respect to this direction of movement. However, it will be carried away from the axis of the work by that portion of the top slide motion that is moving away from the work axis. Thus, the tool is caused to feed out just as if the cross feed were being used. Contours are cut by making that portion of the angular movement that is parallel to the lathe axis, operate at a slower or faster rate than the longitudinal feed rate, while the tool is moving toward or away from the work axis.

QUESTIONS

6-1. What is the origin of the term "engine lathe"?

6-2. Name the principal parts of an engine lathe.

6-3. From what material is the bed of an engine lathe made? Why is this material used?

6-4. What are the back gears of the lathe and what do they do?

6-5. What types of bearings are used to support the lathe headstock?

6-6. Describe three different types of spindle nose designs.

6-7. What is the function of the feed-thread gearbox?

6-8. What is the function of the feed rod? The lead screw?

6-9. Name the principal parts of the carriage.

6-10. What is meant by a direct reading cross-feed dial?

6-11. If a cross-feed dial is not direct reading, how many graduations must it be moved in order to remove .125 inch from the diameter of a part?

6-12. A lathe tool has been positioned to take a cut by moving the cross slide toward the axis of the work. It must then be moved an exact distance away from the work. How is the lost motion of the cross-feed screw overcome in order to make this move of the carriage?

6-13. What is the function of the compound rest?

6-14. What is the rack of the engine lathe, and for what purpose is it used?

6-15. When should the split nut be used?

6-16. How is the body of the tailstock moved perpendicular to the lathe axis?

6-17. Why should large twist drills not be held in the taper of the tailstock spindle?

6-18. What is the micrometer carriage stop and how is it used?

6-19. What is a steady rest and why is it used?

6-20. Describe how a tracer attachment moves the cutting tool in order to cut a square shoulder on a part.

CHAPTER **7**

Single-Point Cutting Tools and Their Performance

Single-point cutting tools are used on engine lathes, shapers, and planers. Boring tools, which are also single-point cutting tools, are used on a large variety of power machine tools to locate and enlarge holes. The performance of these tools has a direct relation to the quality of the work done and the rate at which it is produced. Treatment of this subject will include discussions of: cutting speeds, single-point cutting tool nomenclature, tool shapes, chip formation, chip control, and tool wear. An understanding of these topics will form a necessary foundation for the successful application of these tools in the shop. While the machinist and the toolmaker can only use those machine tools available to them in their respective shops, they will frequently have the responsibility of selecting the best cutting tool and cutting-tool shape available; and, they must know the correct cutting speed and feed with which these tools are to be used.

Hardness and Hardness Testing

Three factors have a very decisive influence on the cutting speed that can be used:

1. The cutting tool material.
2. The material that is to be cut.
3. The metallurgical condition of the material when it is being cut.

One indication of the metallurgical condition of a metal is its hardness; technically this is usually measured by determining its resistance to penetration. The most common hardness-testing machines used in the United States are the Brinell hardness tester and the Rockwell hardness tester. The diameter of the impression made by a steel ball is measured and this measurement is converted to a Brinell hardness number, abbreviated *Bhn*, which defines its hardness. Harder metals have a higher Bhn. There are several Rockwell hardness scales. One which is frequently encountered in measuring the hardness of hardened ferrous metals is the Rockwell C Scale, which is abbreviated R_C. It, too, has the higher numbers indicating higher hardnesses. In general, the Brinell hardness reading is given for softer ferrous metals and the Rockwell reading is given

for hardened ferrous metals, although there is a range in which they overlap.

The strength of a metal increases or decreases roughly in proportion to its hardness. It is not surprising then to discover that harder metals must be cut more slowly than softer metals. The hardness of any piece of soft metal can be increased by *cold drawing* or *cold rolling*. Annealing and normalizing are heat treating processes used to decrease the hardness of steel and cast iron although normalizing may be used to increase the hardness of a piece of steel that has been previously annealed. Most steels, when heated above a specific temperature and then quenched and tempered, will have their hardness increased. *Solution treating and aging* is another heat treatment process which is used to increase the hardness of some metals, particularly aluminum. The details of these processes are a subject in the study of metallurgy; however, the machinist and toolmaker must be aware that the condition of the metals that they are cutting is affected by these processes and that any particular type of metal can have different hardnesses. In fact, it is quite common to have a variation in hardness in a *single* piece of metal. It should now be clear that a cutting speed cannot be given for a specific type of metal without also specifying the hardness. The cutting speeds in Tables 7-1 and 7-3 are, therefore, listed not only for a given metal but also for the different hardnesses at which the metal is commonly cut.

Steels used in the construction of machinery are commonly designated by AISI (American Iron and Steel Institute) numbers. Tool steels are specified by a letter to identify the type, which is followed by a number, such as shown in Table 7-3. Other metals have similar methods of specification. These specifications are in common use in industry and they are used in this book.

Cutting Speed

The cutting speed for all metal cutting operations, including lathe work, is given in feet per minute, which is abbreviated "fpm." In this chapter the primary concern is with the cutting speeds for single-point tools, especially with respect to turning. Recommended cutting speeds for turning are given in Tables 7-1 and 7-3. Similar tables can be found in handbooks on machining as well as in other metal cutting publications. Because of the many factors that affect the cutting speed, and because of their complex nature, the cutting speeds given in these tables must be considered as a starting point only. Actual experience on a given job may indicate the need to modify the recommended speeds. The cutting speeds given in Tables 7-1 and 7-3 are based upon practical workshop experience as well as tests in the machinability laboratory. Before proceeding, the term "tool life" should be defined. *Tool life is the length of time that a cutting tool will cut before becoming dull, or before it must be replaced. It is usually given in minutes.*

The cutting speed for turning depends primarily on the factors listed following, on page 132.

Table 7-1. Recommended Cutting Speeds in Feet per Minute for Turning Ferrous and Nonferrous Metals*

Material	Material Condition	Hardness, Bhn	Cutting Speed, fpm	
			H.S.S.	Carbide
Free Machining, Plain Carbon Steels, (Resulphurized) AISI B1111, B1112, B1113, 1113, 1119, 1212, 1213.	HR, A CD	100 to 150 150 to 200	160 180	500 600
AISI 1108, 1115, 1118, 1120, 1126.	HR, A CD	100 to 150 150 to 200	140 150	450 500
AISI 1132, 1137, 1140, 1145, 1151.	HR, A, N, CD Q & T Q & T Q & T	175 to 225 275 to 325 325 to 375 375 to 425	130 90 50 30	500 250 175 140
Plain Carbon Steels AISI 1012, 1015, 1018, 1019, 1020, 1022, 1024, 1025.	HR, A, N, CD HR, A, N, CD HR, A, N, CD CD	100 to 125 125 to 175 175 to 225 225 to 275	140 120 100 70	500 400 350 300
AISI 1027, 1029, 1030, 1032, 1035, 1037, 1040, 1043, 1045, 1047, 1050.	HR, N, A, CD HR, N, A, CD N, CD, Q & T N, Q & T Q & T Q & T	125 to 175 175 to 225 225 to 275 275 to 325 325 to 375 375 to 425	120 100 70 60 50 40	400 350 300 240 200 175
AISI 1055, 1060, 1065, 1070, 1074, 1080, 1085, 1090, 1095.	HR, N, A, CD HR, N, A, CD N, CD, Q & T N, Q & T Q & T Q & T	125 to 175 175 to 225 225 to 275 275 to 325 325 to 375 375 to 425	100 90 65 55 45 30	375 325 275 225 180 150
Free Machining Alloy Steels (Resulphurized) AISI 3140, 4140, 4150, 8640.	HR, N, A, CD HR, N, A, CD Q & T Q & T Q & T	175 to 200 200 to 250 250 to 300 300 to 375 375 to 425	125 100 70 60 40	450 400 325 225 150
Alloy Steels AISI 1320, 2317, 2512, 2517, 3115, 3120, 3125, 3310, 3316, 4012, 4017, 4023, 4028, 4320, 4615, 4620, 4720, 4815, 4820, 5015, 5020, 5024, 5120, 6118, 6120, 6317, 6325, 6415, 8115, 8615, 8620, 8625, 8720, 8822, 9310, 9315.	HR, A, CD HR, A, N, CD CD, N, Q & T N, Q & T N, Q & T Q & T	150 to 175 175 to 220 220 to 275 275 to 325 325 to 375 375 to 425	110 80 70 60 50 40	400 350 300 250 200 175

* Based upon a feed of .012 inch per revolution and a depth of cut of .125 inch. Symbols under *Material Condition* column, designate: HR-Hot Rolled, A-Annealed, N-Normalized, CD-Cold Drawn or Cold Rolled, Q & T-Quenched and Tempered, AC-As Cast, ST & A-Solution Treated and Aged.

Table 7-1 (cont.). Recommended Cutting Speeds in Feet per Minute for Turning Ferrous and Nonferrous Metals*

Material	Material Condition	Hardness, Bhn	Cutting Speed, fpm	
			H.S.S.	Carbide
AISI 1330, 1335, 1340, 1345, 2330, 2335, 2340, 2345, 3130, 3135, 3140, 3145, 3150, 4030, 4037, 4042, 4047, 4063, 4130, 4135, 4140, 4145, 4150, 4337, 4340, 4640, 5046, 5130, 5135, 5140, 5150, 5160, 50100, 51100, 52100, 6145, 6150, 6240, 6250, 6270, 6342, 6440, 6475, 8630, 8640, 8645, 8650, 8660, 8740, 9255, 9260, 9445, 9840, 9845, 9850.	HR, N, A, CD N, CD, Q & T N, Q & T N, Q & T Q & T	175 to 225 225 to 275 275 to 325 325 to 375 375 to 425	100 80 70 60 40	350 325 275 200 150
Maraging Steels 18% Ni Grade 200. 18% Ni Grade 250. 18% Ni Grade 300.	A	250 to 325	60	300
25% Ni	A	175 to 225	70	250
Free Machining Stainless Steels (Ferritic) 430F, 430F (Se).	A	135 to 185	140	400
(Austenitic) 303, 303(Se), 303F, 303F(Se).	A CD	135 to 185 225 to 275	100 80	350 325
(Martensitic) 416, 416F.	A A, CD Q & T Q & T	135 to 185 185 to 240 275 to 325 375 to 425	140 125 60 30	400 350 250 125
Stainless Steels (Ferritic) 405, 430, 434, 436, 442, 443, 446, 499, 502.	A	135 to 185	100	300
(Austenitic) 201, 202, 301, 302, 304, 304L, 305, 321, 347, 348.	A CD	135 to 185 225 to 275	80 60	250 200
(Martensitic) 403, 410, 420, 501.	A A Q & T Q & T	135 to 175 175 to 225 275 to 325 375 to 425	100 90 50 30	350 300 200 150

* Based upon a feed of .012 inch per revolution and a depth of cut of .125 inch. Symbols under *Material Condition* column, designate: HR-Hot Rolled, A-Annealed, N-Normalized, CD-Cold Drawn or Cold Rolled, Q & T-Quenched and Tempered, AC-As Cast, ST & A-Solution Treated and Aged.

Ch. 7 SINGLE-POINT CUTTING TOOLS AND PERFORMANCE 131

Table 7-1 (cont.). Recommended Cutting Speeds in Feet per Minute for Turning Ferrous and Nonferrous Metals*

Material	Material Condition	Hardness, Bhn	Cutting Speed, fpm	
			H.S.S.	Carbide
(Martensitic) 418, 440A, 440B, 440C.	A Q & T Q & T	225 to 275 275 to 325 375 to 425	50 40 30	200 150 125
Gray Cast-Irons ASTM Class 20	A	110 to 140	120	500
ASTM Class 30	AC	150 to 190	100	325
ASTM Class 40	AC	190 to 220	70	275
ASTM Class 50	AC	220 to 260	50	200
ASTM Class 60	AC, Q & T	250 to 320	30	100
Malleable Irons (Ferritic) 32510, 35018.	...	110 to 160	200	600
(Pearlitic) 45007, 45010, 48004, 50007.	...	160 to 200 200 to 240	90 70	350 300
(Pearlitic) 53004, 60003.	...	200 to 255	70	250
(Pearlitic) 80002.	...	240 to 280	50	200
Aluminum Alloys (Non-Heat-Treated Cast) 13, 43, 85, 108, A108, 112, 113, C113, 132, 138, 212, A214, B214, 218, 319, 360, 380, 384, A612.	AC	...	750	1600
(Heat-Treated Cast) 40E, 122, A132, D132, 142, 195, B195, 220, 319, 355, 356, 750.	ST & A	...	600	1200
(Cold-Drawn Wrought) 1060, 1100, 3003, 3004, 5005, 5052, 5056, 5083, 5086, 5154, 5254, 5357, 5454, 5456, 5652, 5956.	CD	...	600	1200
(Heat-Treated Wrought) 2011, 2014, 2017, 2018, 2024, 2025, 2117, 2218, 4032, 6053, 6061, 6062, 6063, 6151, 6262, 6483, 7075, 7079, 7178.	ST & A	...	600	1200

* Based upon a feed of .012 inch per revolution and a depth of cut of .125 inch. Symbols under *Material Condition* column, designate: HR-Hot Rolled, A-Annealed, N-Normalized, CD-Cold Drawn or Cold Rolled, Q & T-Quenched and Tempered, AC-As Cast, ST & A-Solution Treated and Aged.

132 SINGLE-POINT CUTTING TOOLS AND PERFORMANCE Ch. 7

Table 7-1 (cont.). Recommended Cutting Speeds in Feet per Minute for Turning Ferrous and Nonferrous Metals*

Material	Material Condition	Hardness, Bhn	Cutting Speed, fpm	
			H.S.S	Carbide
Magnesium Alloys (Cast Alloys and Wrought Alloys)	AC, A, ST & A	...	800	2000
Copper Alloys Naval Brass, Red Brass, Yellow Brass, Muntz Metal, Manganese Bronze, Nickel Silver, Aluminum	A CD	300 350	650 700
Commercial Bronze, Phosphor Bronze, Copper Nickel, Oxygen-Free Copper, Electrolytic Tough-Pitch Copper	A CD	100 120	225 250

* Based upon a feed of .012 inch per revolution and a depth of cut of .125 inch. Symbols under *Material Condition* column, designate: HR-Hot Rolled, A-Annealed, N-Normalized, CD-Cold Drawn or Cold Rolled, Q & T-Quenched and Tempered, AC-As Cast, ST & A-Solution Treated and Aged.

1. The type of material from which the cutting tool is made
2. The type of material to be cut, its hardness, and its heat treatment
3. The condition of the surface of the material through which the cut is to be made
4. The shape of the cutting tool, particularly with respect to certain critical angles
5. The desired tool life of the cutting tool
6. The type of cutting fluid used, if any
7. The type of cut, whether a roughing or a finishing cut
8. The feed and the depth of cut to be used.

Many other factors also influence the speed at which metal can be cut. The surface of castings and hot rolled bars is often hard and abrasive; this results in rapid tool wear and may require a reduction in the cutting speed recommended in the tables. Sometimes castings which have not been heat treated have hard spots. Cutting fluids can sometimes be used to allow a higher cutting speed to be used. They are particularly effective in improving the surface finish obtained when the lower range of cutting speeds is used. In some cases the cutting fluid will react with the surface of the cutting tool, causing it to wear more rapidly; however, this is not objectionable when a good surface finish is desired. The shape of the cutting tool has an influence on the cutting speed for a given tool life, as will be discussed further on.

The tool life and the cutting speed are directly related to each other. When the cutting speed is reduced, the tool life will always increase. This increase in tool life will always be proportionately greater than the de-

Ch. 7 SINGLE-POINT CUTTING TOOLS AND PERFORMANCE

crease in cutting speed. Conversely, increasing the cutting speed will cause a proportionately greater decrease in the tool life. This is a very important performance characteristic of metal-cutting tools. *The practical implication of this phenomenon is that most metals can be cut successfully using a fairly wide range of cutting speeds.* The exceptions are some of the high-temperature alloys and metals having a very high hardness. In most cases, however, the objective is to select a cutting speed that will cut metal at a fast rate while at the same time giving a reasonable tool life. Generally, an extremely long tool life is undesirable if it is obtained at the expense of an uneconomical metal-removal rate.

More tool wear can be permitted in taking roughing cuts because the surface finish requirements are usually not as great as when finish turning. A finish-turning tool must not be allowed to wear as much as a roughing tool, in order to obtain a good surface finish. Increasing the cutting speed beyond a certain critical cutting speed results in a very great improvement in the surface finish; however, not all cutting-tool materials can cut at this speed without failing. Cemented-carbide and ceramic cutting-tool materials can cut successfully above this speed. They should always be used at cutting speeds that are fast enough to give a good surface finish. Industrial diamonds are used as single-point tools to finish machine hard-to-finish materials such as soft aluminum.

The feed rate and the depth of cut used also have a pronounced effect on the cutting speed. The feed rate has a greater influence than the depth of cut. This can be seen by noting the feed and the depth-of-cut correction factors given in Table 7-2. This table is used in conjunction with Table 7-1. The recommended cutting speeds in Table 7-1 are for a feed of .012 inch per revolution and a depth-of-cut of .125 inch. When a different feed or depth of cut is to be used, the recommended cutting speed in Table 7-1 should be multiplied by the correction factors given in

Table 7-2. Cutting Speed Correction Factors for Feed and Depth of Cut When Turning (To be used with Table 7-1)

Feed, ipr	Correction Factor, F_f	Depth of Cut, Inch	Correction Factor, F_d
0.003	2.00	0.005	1.80
0.005	1.69	0.010	1.50
0.008	1.27	0.020	1.40
0.010	1.12	0.031	1.30
0.012	1.00	0.062	1.15
0.015	0.87	0.094	1.05
0.018	0.78	0.125	1.00
0.020	0.74	0.150	0.96
0.022	0.70	0.200	0.91
0.025	0.64	0.250	0.87
0.028	0.61	0.312	0.83
0.030	0.58	0.375	0.80
0.036	0.52	0.500	0.76
0.040	0.48	0.625	0.72
0.050	0.42	0.750	0.70
0.060	0.38	1.000	0.66

Table 7-2. The correction factors given in Table 7-2 are based on extensive test results and are reasonably accurate. They should not, however, be used for the recommended cutting speeds given in Table 7-3 as their validity for tool steels is uncertain.

From the discussion of the factors affecting cutting speeds, the difficulty of giving precise recommendations covering every situation is evident. This should not detract from the usefulness of cutting-speed data as these data represent the best currently available information. It should

Table 7-3. Recommended Cutting Speeds for Turning Tool Steels (fpm)*

Tool Steel	Material Condition	Hardness	Cutting Speed, fpm	
			H.S.S.	Carbide
Water Hardening W1, W2, W4, W5.	A	150 to 200 Bhn	110	350
Cold Work A7, D1, D2, D3, D4, D5, D7.	A	200 to 250 Bhn	40	160
A2, A3, A4, A5, A6, A8, A10, O1, O2, O6, O7.	A	200 to 250 Bhn	70	250
Hot Work H10, H11, H12, H13, H14, H16, H19.	A A Q & T Q & T Q & T Q & T Q & T	150 to 200 Bhn 200 to 250 Bhn 325 to 375 Bhn 48 to 50 R_c 50 to 52 R_c 52 to 54 R_c 54 to 56 R_c	90 75 50 20 15	300 260 200 95 80 60 40
H20, H21, H22, H23, H24, H25, H41, H42, H43.	A A	150 to 200 Bhn 200 to 250 Bhn	60 50	250 225
Shock Resisting Steel S1, S2, S4, S5, S6, S7.	A	175 to 225 Bhn	80	300
Mold P1, P2, P3, P4, P5, P6.	A	100 to 150 Bhn	110	350
P20, P21.	A	150 to 200 Bhn	90	300
Special Purpose L1, L2, L3, L6, L7.	A	150 to 200 Bhn	90	300
F1, F2, F3.	A	200 to 250 Bhn	70	225
High-Speed Steel M1, M2, M6, M10, T1, T2, T6, T7.	A	200 to 250 Bhn	60	200
M3 Class 1, M4, M7, M30, M33, M34, M35, M36, M41, M42, M43, M44, T4, T5, T8.	A	225 to 275 Bhn	50	175
M3 Class 2, M15, T9, T15.	A	225 to 275 Bhn	40	140

* Symbols under *Material Condition* column, designate: A-Annealed, Q & T-Quenched and Tempered.

Ch. 7 SINGLE-POINT CUTTING TOOLS AND PERFORMANCE 135

be used as a point of departure which may be modified, if required, as experience is gained on any particular job. One additional principle of metal-cutting tool performance that should be kept in mind when adjusting the cutting conditions is that *the tool life of the cutting tool is most affected by the cutting speed, secondly by the feed, and least of all by the depth of cut.* Thus, again, if longer tool life is desired, it is best to reduce the cutting speed.

The formula for calculating the cutting speed, using Tables 7-1 and 7-2, is given below, along with the formula for calculating the spindle speed at which the lathe should operate:

$$V = V_o F_f F_d \tag{7-1}$$

$$N = \frac{12 V}{\pi D} \tag{7-2a}$$

$$V = \frac{\pi D N}{12} \tag{7-2b}$$

where:

V = Cutting speed in feet per minute. When used with Tables 7-1 and 7-2 this speed is corrected for feed and depth of cut.
V_o = Original cutting speed, in feet per minute, as given in Table 7-1, for a feed of .012 inch per revolution and a depth of cut of .125 inch.
F_f = Feed Correction factor given in Table 7-2.
F_d = Depth of cut correction factor given in Table 7-2.
N = Lathe-spindle speed in revolutions per minute.
D = Diameter, in inches. This is the diameter to be turned, not the finish-turned diameter.
π = 3.14 (pi)

Example 7-1:

Determine the cutting speed and spindle speed that should be used to turn a 3-inch-diameter bar of AISI 1040 steel which has a hardness of 180 Bhn. The feed to be used is .020 inch per revolution and the depth of cut is .250 inch. The cutting-tool material is high-speed steel.

$$V_o = 100 \text{ (from Table 7-1)}$$

$$F_f = .74 \text{ (from Table 7-2)}$$

$$F_d = .87 \text{ (from Table 7-2)}$$

$$V = V_o F_f F_d$$
$$= 100 \times .74 \times .87$$

$$V = 64.4 \text{ fpm}$$

$$N = \frac{12\ V}{\pi D}$$

$$= \frac{12 \times 64.4}{3.14 \times 3}$$

$$N = 82 \text{ rpm}$$

Example 7-2:

Calculate the lathe-spindle speed when a bar of AISI 4140 steel, with a Bhn of 220 is to be turned using a cemented carbide cutting tool. The diameter of the bar of steel is 4 inches. The feed rate to be used is .018 inch per revolution and the depth of cut is .125 inch.

$$V_o = 300 \text{ (from Table 7-1)}$$

$$F_f = .78 \text{ (from Table 7-2)}$$

$$F_d = 1.00 \text{ (from Table 7-2)}$$

$$V = V_o F_f F_d$$
$$= 300 \times .78 \times 1.00$$

$$V = 234 \text{ fpm}$$

$$N = \frac{12\ V}{\pi D} = \frac{12 \times 234}{3.14 \times 4}$$

$$N = 224 \text{ rpm}$$

Example 7-3:

Calculate the cutting speed that is being used when a 5-inch-diameter hole is being bored, using a high-speed-steel cutting tool. The spindle speed of the lathe is 76 rpm.

(Note: This is a typical application of Formula 7-2b.)

$$V = \frac{\pi\ DN}{12}$$

$$= \frac{3.14 \times 5 \times 76}{12}$$

$$V = 99.5 \text{ fpm}$$

Cutting Tool Materials

There are six basic types of metal-cutting tool materials. They are plain carbon tool steel, high-speed steel, cast nonferrous alloys, cemented carbides, ceramics, and diamonds. There are numerous grades of each of these materials and in most instances the difference between these grades is significant. The cutting tool materials must be hardened to high-hardness value. Preferably they should harden deeply. The ability of a steel to harden deeply is called its hardenability. They should have good

Ch. 7 SINGLE-POINT CUTTING TOOLS AND PERFORMANCE 137

red hardness, or hot hardness—a property which relates to their ability to retain their hardness at the high temperatures encountered by their cutting edge in the act of cutting metal. They should have good abrasion resistance and good crater resistance. Crater resistance refers to the resistance to the formation of a small crater on the face of the cutting tool.

Plain Carbon Tool Steel

Plain carbon tool steels have .90 to 1.30 per cent carbon; the remainder being iron. Some carbon tool steels contain a small amount of vanadium and chromium to increase their toughness and depth of hardness. Carbon tool steel is seldom used to make single-point cutting tools because the cutting speed with which it can be used is relatively very low. The cutting speed that can be used for plain carbon tool steel is approximately one-half of the cutting speed for high-speed steel. It is used for making some hand reamers, and less expensive drills and taps.

High-Speed Steel

High-speed steel, of which there are many grades, is the most extensively used cutting-tool material. The chemical compositions of the most frequently used high speed steels are given in Table 7-4. The most widely

Table 7-4. Chemical Composition of Frequently Used High-Speed Steels*

AISI Designation	Nominal Chemical Composition, per cent					
	C	W	Mo	Cr	V	Co
M-1	0.80	1.50	8.00	4.00	1.00	...
M-2	0.85	6.00	5.00	4.00	2.00	...
T-1	0.70	18.00	...	4.00	1.00	
T-4	0.75	18.00	...	4.00	1.00	5.00
T-6	0.80	20.00	...	4.50	1.50	12.00
T-15	1.50	12.00	...	4.00	5.00	5.00

* C-Carbon, W-Tungsten, Mo-Molybdenum, Cr-Chromium, V-Vanadium, Co-Cobalt

used type steel is the M-2. It is used to make single-point cutting tools as well as all types of multiple pointed tools, such as drills, reamers, taps, and milling cutters. The type T-1, is the original high-speed steel. It is still used to make single-point cutting tools. The addition of cobalt to high-speed steel increases its red hardness, making it particularly adaptable when taking roughing cuts. The T-4, cobalt high-speed steel, is recommended for machining cast iron, and the T-6 is recommended for machining hard, scaly, or gritty materials. The T-15, high-speed steel is one of a class of "Super High-Speed Steels" that have been designed to machine difficult-to-machine materials such as stainless steels, high-temperature alloys, and refractory metals. The T-15, high-speed steel is very hard (up to 70 R_C) and must be ground slowly with a very light grinding pressure. The cutting speed recommended for high-speed steel is given in Tables 7-1 and 7-3.

Cemented Carbides

Cemented carbides are harder than high-speed steels and can operate at considerably higher cutting speeds. They are made from powdered metals that have been compressed and sintered at temperatures of approximately 2400° F. They are not porous—being as compact and dense as other metals—and have excellent wear-resistance properties. They are, however, more brittle and more costly than high-speed steel. For this reason cemented carbides are used as small tips or inserts on the end of steel shanks or cutter bodies.

Failure to select the correct carbide grade for each application may result in very poor performance; hence, it is best to seek a carbide manufacturer's recommendation. Only a few very broad recommendations are made here. The cutting speeds for carbides given in Tables 7-1 and 7-3 are based on the assumption that the best grade of carbide for the particular metal is being used.

Straight tungsten carbides contain only tungsten carbide and cobalt. Approximately **94 to 97** per cent of the composition is tungsten carbide, the remainder is cobalt. The higher cobalt-bearing carbides are tougher and more shock resistant; recommended for taking heavy roughing cuts. The higher tungsten carbide bearing carbides are harder, more wear resistant, and are recommended for light-duty boring, and finishing cuts that require good, abrasive wear resistance. Straight tungsten carbides are recommended for machining gray cast-iron, ferritic malleable iron, austenitic, stainless steels, high temperature alloys, copper, brass, bronze, aluminum, zinc die-cast alloys, and plastics. They should not be used to machine plain carbon and alloy steels as these materials will cause the straight tungsten carbides to crater and wear out rapidly.

Crater-resistant carbides are made by adding titanium carbide and tantalum carbide to the tungsten carbide and cobalt. The harder grades of the crater-resistant carbides are used for taking medium and light duty cuts and the softer carbides are used for the heavy duty cuts. These carbides are recommended for machining all grades of plain carbon and alloy steels, alloy cast-iron, pearlitic malleable iron, nodular iron, Monel metal, martensitic stainless steel, ferritic stainless steel, and the tool steels.

A third class of cemented carbides is made from titanium carbide and a small quantity of either nickel or molybdenum which acts as a filler or binder between the titanium carbide grains. They have excellent wear and crater resistance, but their shock resistance is somewhat low. Titanium carbides are recommended for taking light finishing and precision boring cuts, although in some circumstances, they have been successfully used to take medium duty cuts. An excellent surface finish can be obtained with this class of carbide.

Ceramic Materials

Ceramic cutting tool materials are made by sintering together into a dense structure, fine grains of aluminum oxide. Sometimes, small amounts

Ch. 7 SINGLE-POINT CUTTING TOOLS AND PERFORMANCE

of titanium oxide, magnesium oxide, or chromium oxide are added. They are harder than cemented carbides and have poor thermal conductivity. Thus, they do not readily absorb the heat generated in forming the chip. A good tool life has been attained when cutting AISI 1040 steel at a cutting speed of 18,000 fpm on a special, high-speed machinability lathe. Ordinary machine tools, however, are not designed to obtain such high speeds. Ceramic cutting tools are recommended for turning plain carbon steels, alloy steels, pearlitic malleable iron, gray cast-iron, nodular iron, and stainless steel.

Cast Nonferrous Alloys

Cast nonferrous alloys can be used for cutting at speeds about 20 to 50 per cent higher than high-speed steels. They are not as tough, but are more wear resistant than high-speed steel. They cannot cut as fast as cemented carbides and their application is limited. The composition of these alloys is approximately 40 to 53 per cent cobalt, 30 to 33 per cent chromium, and 14 to 22 per cent tungsten.

Industrial Diamonds

Industrial diamonds are used as single-point cutting tools to obtain an excellent surface finish on hard-to-finish materials such as aluminum, magnesium, and bearing metals. Their ability to produce a good surface finish is outstanding. They are used for taking light-finish turning and boring cuts using a depth of cut of .003 to .030 inch and a feed of .0008 to .005 inch per revolution.

Single-Point Cutting Tool Nomenclature

The nomenclature of a single-point cutting tool is shown in Fig. 7-1. Most cutting tools have two cutting edges; a side-cutting edge and end-cutting edge (see View A, Fig. 7-1). When doing ordinary turning the cutting is done with the side-cutting edge. The nose is that portion of the cutting edge which serves to connect the two cutting edges. It is a very critical part of the cutting edge because it produces the finished surface in turning. The flank of the tool is the surface on the side of the tool below the cutting edges. It must be relieved in order to permit the cutting edges to penetrate into the metal being cut. If the flanks below the side-cutting edge, nose, and end-cutting edge were not relieved, the tool would rub and could not penetrate into the workpiece. The face of the cutting tool provides a surface over which the chip that is formed by the cut slides. It also has a great influence in determining the angle of the shear plane on which the grains of the metal are deformed in order to form the chip. This will be described later in this chapter. The metal remaining in back of the cutting edge is the shank. It supports the cutting edge and provides a surface upon which the cutting tool can rest and be held.

The cutting tool in Fig. 7-1 is a right-hand tool. A right-hand cutting tool can be identified by having the side-cutting edge on the left when viewed from the shank. A left-hand cutting tool has the side-cutting edge

140 SINGLE-POINT CUTTING TOOLS AND PERFORMANCE Ch. 7

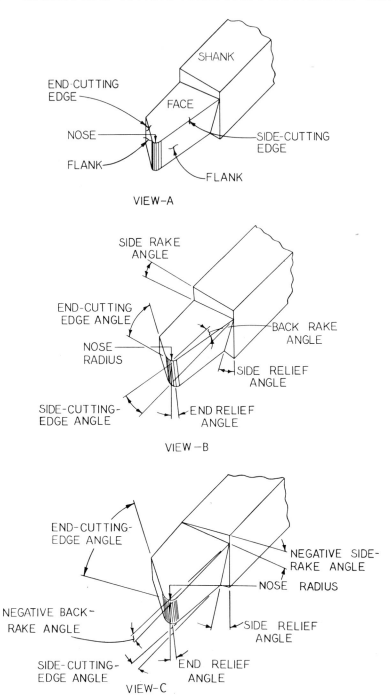

Fig. 7-1. The nomenclature of the single-point cutting tool.

Ch. 7 SINGLE-POINT CUTTING TOOLS AND PERFORMANCE 141

on the right side when viewed from the shank. The tool shown is also classified as a side-cutting-edge tool because it is provided with a side-cutting edge and because it is intended that most of the cutting will be done with the side-cutting edge. End-cutting-edge tools (see G, H, and J, Fig. 7-9) are intended to cut exclusively with their end-cutting edges and are made without a side-cutting edge.

When specifying a single-point cutting tool, it is necessary to specify the width and height of the shank. In addition, it is necessary to specify the radius of the nose and the six principal tool angles which are: 1. end relief angle; 2. side relief angle; 3. back rake angle; 4. side rake angle; 5. end-cutting-edge angle; and 6. side-cutting-edge angle. These angles are all shown in View B, Fig. 7-1.

Relief Angles

The side relief angle provides the relief for the side-cutting edge in order to allow it to penetrate into the workpiece so that it can form the chip. The end relief angle allows the end-cutting edge to penetrate into the workpiece. The relief angle below the nose is a blend of both the end relief and the side relief angles. Usually, both relief angles are equal although this is not always the case. The size of the relief angles has a very pronounced effect on the performance of the cutting tool; if they are too small they will cause a decided decrease in the tool life; if too large, the support below the cutting edge will be weakened to the extent that the cutting edge may break or chip. For average work, a relief angle of 10 degrees is recommended. For high-speed-steel cutting tools the relief angle should be in the range of 8 to 16 degrees. Harder materials may require a smaller relief angle while softer materials are more successfully cut when a larger relief angle is used. The relief angle for cemented-carbide cutting tools should be in the range of 5 to 12 degrees. Since carbides are harder and more brittle than high-speed steel, a better support under the cutting edge is required.

Rake Angles

The slope of the face of the cutting tool is determined by the back-rake and side-rake angles. The side-rake angle is measured perpendicular to the side-cutting edge and the back-rake angle is measured parallel to the side-cutting edge. The rake angles may be positive, as shown at B in Fig. 7-1, or may be negative, as shown at C. Note that the positive rake angles cause the face of the cutting tool to slant downward when moving away from the cutting edges, and that negative rake angles cause this surface to slant upward when moving away from the cutting edges. Positive side-rake angles are generally preferred because less cutting force is required in order to take a given size cut as compared to a tool with a negative side-rake angle.

The cutting force required to take a given cut will increase approximately one per cent per degree that the size of the positive rake angle is decreased, or that the size of the negative side-rake angle is increased.

For example, if a tool with a 5-degree, positive side-rake angle is replaced by one with a 5-degree negative side-rake angle, the force required to take the cut will be increased approximately ten per cent. The back-rake angle does not have as much effect on the cutting force. The positive side-rake angle tool will generate less heat and will have a longer tool life than the negative side-rake angle cutting tool. The side-rake angle on most cutting tools should be made with as large a positive rake-angle as possible, up to a certain limit, for if it is made too large, the cutting edge will be weakened and will readily fail. The advantage of having a negative side-rake angle, and a negative back-rake angle, is that the cutting edge is strengthened; thereby enabling it to withstand more severe cutting loads. This is an advantage when cutting very hard and tough materials and when taking interrupted cuts. Negative rake angles are often preferred on disposable or throw-away carbide cutting tools (see Fig. 7-5) because relief angles do not have to be ground on the inserts; also, the edges on both the top and the bottom of the inserts can be used as cutting edges. This provides twice as many cutting edges as a comparable, positive rake cutting tool which has cutting edges avail-

Table 7-5. Recommended Rake Angles

Material	Hardness, Bhn	High-Speed Steel		Carbide	
		Back Rake Angle, deg.	Side Rake Angle, deg.	Back Rake Angle, deg.	Side Rake Angle, deg.
Plain Carbon Steel	100 to 200	5 to 10	10 to 20	0 to 5	7 to 15
	200 to 300	5 to 7	8 to 12	0 to 5	5 to 8
	300 to 400	0 to 5	5 to 10	−5 to 0	3 to 5
	400 to 500	−5 to 0	−5 to 0	−8 to 0	−6 to 0
Alloy Steel	100 to 200	5 to 10	10 to 16	0 to 5	5 to 15
	200 to 300	5 to 7	8 to 12	0 to 5	5 to 8
	300 to 400	0 to 5	5 to 10	−5 to 0	3 to 5
	400 to 500	−5 to 0	−5 to 0	−6 to 0	−6 to 0
Aluminum					
Non Heat-Treated	...	10 to 20	30 to 35	0 to 15	15 to 30
Heat Treated	...	5 to 12	15 to 20	0 to 5	8 to 15
Magnesium	...	5 to 10	10 to 20	0 to 5	10 to 20
Stainless Steel					
Ferritic	130 to 190	5 to 7	8 to 10	0 to 5	5 to 7
Austenitic	130 to 190	5 to 7	8 to 12	0 to 5	5 to 7
Martensitic	130 to 220	0 to 5	5 to 8	−5 to 5	6 to 15
Gray Cast-Iron	100 to 200	5 to 10	10 to 15	0 to 5	6 to 15
	200 to 300	5 to 7	5 to 10	0 to 5	5 to 7
	300 to 400	−5 to 0	−5 to 0	−6 to 0	−6 to 0
Malleable Iron					
Ferritic	110 to 160	5 to 15	12 to 20	0 to 10	7 to 15
Pearlitic	160 to 200	5 to 8	10 to 12	0 to 5	5 to 8
	200 to 280	0 to 5	−5 to 8	−5 to 5	−5 to 8
Brass-Free Cutting	...	−5 to 5	0 to 10
Brass					
Red, Yellow, Naval, Manganese Bronze, Nickel Silver	...	−5 to 5	−5 to 5	−5 to 5	−5 to 5
Hard Phosphor Bronze	...	−5 to 0	−6 to 3	−5 to 0	−5 to 5

Ch. 7 SINGLE-POINT CUTTING TOOLS AND PERFORMANCE 143

able on only one face of the insert. The rake angles recommended for cutting different work materials are given in Table 7-5.

End-Cutting-Edge Angle

The end-cutting-edge angle is the amount that the end-cutting edge slopes away from the nose of the tool, so that it will clear the finished surface on the workpiece, when cutting with the side-cutting edge. The size of this angle is very important, particularly when cutting materials that tend to form a large crater on the face of the tool. This crater will then tend to enlarge toward the end-cutting edge where it will eventually break through and cause the tool to fail. This is shown at B, Fig. 7-18. When severe cratering occurs, the size of the end-cutting edge should be limited to 8 to 15 degrees; otherwise, it can be made as large as 45 degrees without an adverse effect on the performance of the tool.

Side-Cutting Edge and Lead Angles

Since the chip will flow approximately perpendicular to the side-cutting edge, the side-cutting-edge angle, thereby, influences the direction of the chip flow. It also affects the thickness of the chip and the direction of the longitudinal component of the cutting force. These, however, are more affected by the lead angle. An understanding of the relationship of the lead angle to the side-cutting-edge angle is necessary in order to understand how single-point cutting tools perform.

The lead angle is defined as the angle at which the side-cutting edge is positioned, with respect to a plane that is perpendicular to the axis of the workpiece. (See Fig. 7-2.) When the cutting tool is positioned perpendicular to the workpiece, the lead angle is equal to the side-cutting-edge angle as shown at B. In Fig. 7-3, the tool at A, has a zero degree lead angle and the tool at B, has a larger lead angle. The feed, or distance that the cutting tool advances per revolution, and the depth of cut are

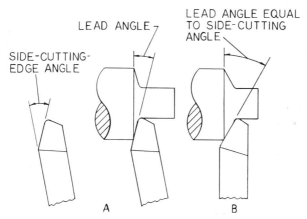

Fig. 7-2. Definition of the lead angle.

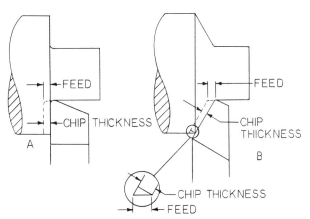

Fig. 7-3. The effect of the lead angle on the chip thickness.

equal, in both cases. The chip thickness is measured perpendicular to the side-cutting edge of the tool. At A, the chip thickness is equal to the feed; while at B, the chip thickness is less than the feed. In each case the same amount of metal is removed—in the form of a chip—for each revolution of the work; however, the chip at B, is longer and thinner. Thus, the effect of the lead angles is to produce a longer and thinner chip, which can be considered the equivalent of increasing the depth of cut and decreasing the feed rate. A study of the effect of feed on the cutting speed, in Table 7-2, will show that the thinner chip, or lower feed rate, will permit the cutting speed to be increased; or, if the cutting speed remains the same, the feed can be increased. In either case, an increase in the lead angle can bring about an increase in the production rate that can be attained. The effect of the depth of cut is relatively small, and can usually be ignored.

There is, however, an adverse effect caused by a large lead angle. This is shown in Fig. 7-4. A force is required to feed the cutting tool into the workpiece; this force is called the "longitudinal" force. It is perpendicular to the side-cutting edge. Although shown in Fig. 7-4, as reacting against the work, it reacts with an equal magnitude against the cutting tool. At A, this force is parallel to the axis of the workpiece and there is very little tendency to cause it to bend the work. At B, this force is directed so that it will cross the workpiece axis which will cause the workpiece to bend away from the tool. The elasticity in the workpiece material will cause it to spring back into its original position. When this cycle is repeated at certain frequencies a very noticeable vibration, called chatter, will set in. Chatter impairs the surface finish on the work and causes a rapid breakdown of the cutting edge as well. In all instances it must be eliminated. It can be caused by many factors, such as: loose headstock-spindle bearings, a loose fit on the cross slide, too much overhang of the cutting tool, or a very dull cutting tool. However, the most frequent cause

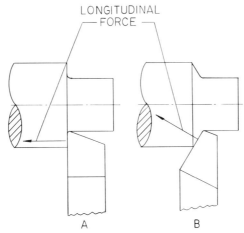

Fig. 7-4. The direction of the reaction of the longitudinal cutting force is perpendicular to the side-cutting edge. An equal and opposite force that is not shown reacts against the cutting edge.

is a lead angle that is too large; especially when turning a long and slender workpiece. The first step to take is to examine the lead angle used and to reduce it if it is too large. In a great number of instances, this will eliminate the chatter completely. When boring with a cutting tool having a large lead angle, the reaction of the longitudinal force against the cutting edge will deflect the boring bar, thus causing chatter or some difficulty in boring the hole to a desired dimension. For this reason the lead angle on boring tools should be kept small. When boring either very deep or small diameter holes, it is best to use a zero degree lead angle.

Nose Radius

The nose is an extension of the cutting edges. Its point must be formed into a radius in order to prevent the formation of a sharp threadlike groove on the surface of the work. When taking a cut, the chip thickness of the nose is equal to the chip thickness of the side-cutting edge at the point where they join. At the point where the nose is tangent to the workpiece, the chip thickness becomes zero in magnitude. Thus, the chip becomes increasingly thinner as the point of tangency between the nose and the work is approached. If the nose radius is large, there will be a large area where a very thin chip is formed. When this occurs the cutting edge may fail to penetrate the work in this region, thus causing it to rub. This is very apt to occur if the cutting edge is dull. Should the cutting edge fail to penetrate the work, chatter may be set up or the tool will get dull rapidly, in this region. Thus, we have another cause of chatter, a nose radius that is too large. Hence, reducing the size of the nose radius will reduce the tendency to chatter. When finish turning, the nose

146 SINGLE-POINT CUTTING TOOLS AND PERFORMANCE Ch. 7

radius must be made large enough to prevent any formation of thread-like grooves that have the form of the nose. If the size of the nose radius cannot be increased, the feed must be reduced to obtain a good surface finish. The size of the nose radius is then dependent upon the feed rate, the surface finish required on the work, and the requirement to prevent the occurrence of chatter. It should be pointed out that the life of the cutting tool can be adversely affected by a nose radius that is either too large or too small. For average conditions when turning, a nose radius of $\frac{1}{64}$ to $\frac{1}{8}$ inch is recommended, although larger nose radii can sometimes be used successfully.

Tool Holders

In most cases the cutting tool is clamped in a tool holder; however, single-point cutting tools are sometimes clamped directly on the machine. The tool holder is made from a less expensive steel than the cutting tool. It is heat treated to increase its strength and to resist the penetration of the screws which clamp it in place. Although there is a great variety of tool holders, only representative examples of commonly used designs are illustrated here. One common but very effective tool holder for holding high-speed-steel tool bits, is shown in A, Fig. 7-5. Cemented

Courtesy of Armstrong Brothers Tool Company
Courtesy of Metallurgical Products Dept., General Electric Co.
Courtesy of Kennametal Inc.

Fig. 7-5. A (Upper left) Tool holder for high-speed steel tool bits. B (Upper right) A cemented-carbide cutting tool with a brazed-on insert. C & D (Lower left and right) Throwaway-insert, carbide tool-holder assembly.

Ch. 7 SINGLE-POINT CUTTING TOOLS AND PERFORMANCE 147

carbide tips are sometimes brazed on steel shanks as in B, Fig. 7-5. At C, Fig. 7-5, a clamped throwaway, or disposable, carbide-insert tool holder is shown. The construction of this holder is shown at D. The clamp presses down on a carbide plate-type chip breaker which, in turn, holds the carbide insert in place. The carbide insert is the actual cutting tool. When cutting metals that form a discontinuous chip, the chip breaker need not be used, and the clamp presses directly on the carbide insert, or cutting tool. The insert rests on a carbide seat which is held by a screw in the bottom of the pocket of the tool holder. This carbide insert has

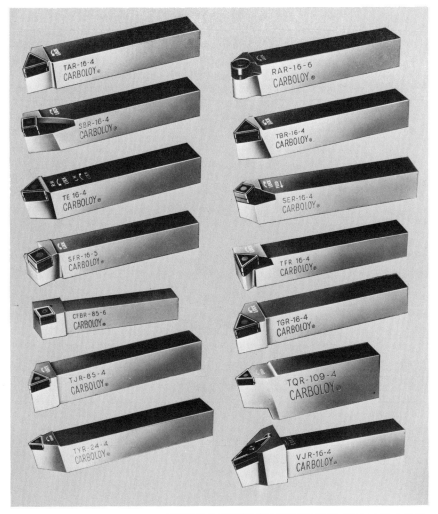

Courtesy of Metallurgical Products Dept., General Electric Co.

Fig. 7-6. Carb-O-Lock, disposable-insert tool holders for cemented-carbide inserts. Inserts shown have pressed-in chip breakers.

eight cutting edges: four on the top face and four more on the bottom. After one cutting edge has been used up the clamp is loosened and the insert is indexed to bring another cutting edge into position. After all of the cutting edges have been used up it is thrown away. The tool holder shown is a negative-rake tool holder. Positive rake, disposable-insert tool holders are also made. The inserts for the positive-rake tool holders must have a relief angle ground on their sides; therefore, only the cutting edges on the top face can be used. The inserts used by both positive and negative-rake tool holders may be shaped square, triangular, round, and diamond, as shown in Fig. 7-6. These inserts have a hole in the center and are held in place by an eccentric cam. A chip-breaker groove is formed in these inserts when they are pressed prior to sintering. Inserts without the chip-breaker groove can also be obtained for cutting metals that form a discontinuous chip. A tool holder and a boring bar for holding carbide inserts that are used to cut screw threads, are shown in Fig. 7-7. Two types of boring bars for holding high-speed-steel tool bits are shown in Fig. 7-8.

Tool Shapes

Many different tool shapes have been successfully used for turning operations but only a few of the basic shapes can be shown here. Regardless of the shape of the single-point tool, it must have the optimum tool angles that will allow it to penetrate the metal and cut efficiently. This section will deal only with the *shape* of the tool, since the tool angles have been described in a previous section.

Courtesy of Kennametal Inc.

Fig. 7-7. Throwaway-insert tool holder and boring bar for thread cutting.

Ch. 7 SINGLE-POINT CUTTING TOOLS AND PERFORMANCE 149

Courtesy of Armstrong Brothers Tool Company

Fig. 7-8. Boring bars for high-speed tool bits.

Some typical shapes for high-speed-steel cutting tools are shown in Fig. 7-9. The tool at A is used to turn the cylindrical surface and to form a square shoulder at the end of the cut. The tool at B is a turning tool having a lead angle. It could be turned so that the cutting edge has a zero degree lead angle, thereby enabling it to cut a square shoulder at the end of the turning operation. The turning tool at C has a large-nose radius that will produce a good surface finish when taking either heavy or light cuts. The curved cutting edge is, in effect, a large lead angle. It tends to direct the flow of the chips away from the finished surface on the work. The tool at D has the nose leading the side-cutting edge. In this position, the cutting tool is set to take a light finishing cut on the cylindrical surface and on the face of the shoulder. This tool position, having the nose leading the cutting edge, is frequently used to turn shoulders and contours on numerically controlled, and tracer controlled, lathes. The tool at E is a form tool which reproduces on the work, the form ground on the cutting edge. An end-facing tool is shown at F. It is used to face the ends of workpieces that are held between the centers in the lathe. Tool G is called a necking, or grooving, tool. It is used to cut grooves as shown. A cut-off tool is shown at H. This tool is used to cut off stock or finished parts that are held in a chuck. A left-hand turning tool (one that cuts from left to right in conventional turning) is shown at I, facing the end of a part that is held in a chuck. A different type of end-facing tool is shown at J. Cutting is done with the end-cutting edge which should be honed to have a flat surface from .003- to .010-inch wide. The "hook" that is ground in the tool is equivalent to the side-rake angle on turning tools. The side of the tool which is adjacent to the

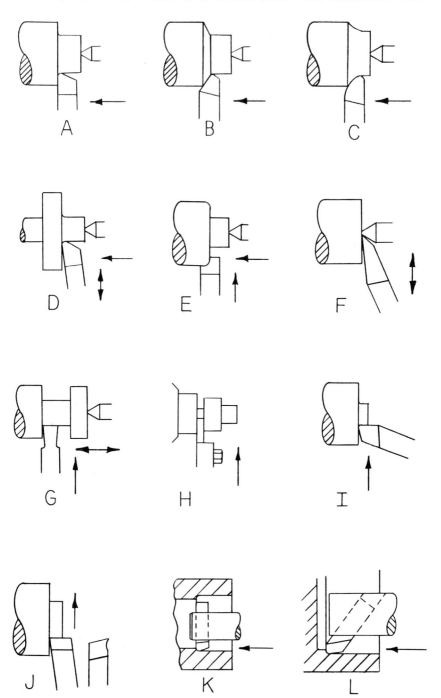

Fig. 7-9. Typical tool shapes and applications for high-speed-steel tool bits.

Ch. 7 SINGLE-POINT CUTTING TOOLS AND PERFORMANCE 151

Fig. 7-10. Throwaway-insert tool holders and their application.

shoulder should have a 5- to 8-degree relief angle. A boring tool is shown at K. Another boring tool, at L, permits cutting up to the face of large internal shoulders. The shoulder can also be cut with this boring tool.

The shape of disposable insert, cemented-carbide cutting tools is determined by the shape of the insert and the manner in which the insert is held in the tool holder. These have been standardized by the carbide manufacturers. They are designated by a letter code, as shown in Fig. 7-10, which shows how these tools are applied. For example, the A-style tool has a zero-degree side-cutting-edge angle and is used to turn, face, or bore to a square shoulder. The B-style has a 15-degree side-cutting-edge angle used for rough turning, facing, or boring, where a square shoulder is not required. The C-style tool is used for grooving, chamfering, or plunge cutting. The D-style has a 45-degree side-cutting-edge angle. It is used for rough turning (when the depth of cut is not excessively deep) and for finish turning. It provides the advantages of a large lead angle; however, it cannot be used to take heavy cuts when chatter occurs. The E-style tool is used for threading and making "V" grooves. The F-style tool is used for facing and turning, as shown. When it is necessary to turn close to a chuck or a shoulder, the G-style tool is recommended. The carbide tools shown in Fig. 7-6 have the same classification. The diamond-shaped tool, which is the lowest in the right-side

Fig. 7-11. A. Combination rough- and finish-turning tool. B. Combination rough- and finish-turning tool with lead angle. C. Broadnose finish-turning tool.

column of tools, shown in Fig. 7-6, is classified as a J-style tool. It is used to cut contours on numerically controlled, and tracer controlled, lathes.

Some very useful single-point tool shapes are shown in Fig. 7-11. The tool at A, is a combination "rough and finish" turning tool. A small flat is ground on the end-cutting edge, which is used to take a very light finishing or scraping cut, while the side-cutting edge is used to take a heavier cut. The width of the flat should be from $\frac{1}{16}$ to $\frac{3}{16}$ inch. The side rake angle of this tool must be zero degrees. It may be given a positive back rake angle up to 10 degrees. A similar tool, having a 45-degree, side-cutting-edge angle is shown at B. Either tool can be made from high-speed steel or from cemented carbide. The flat of each of these tools can be used to take a light finishing cut when desired, although, primarily, they are designed to take combined roughing and finishing cuts.

A broad-nose finishing tool is shown at C, Fig. 7-11. When properly ground, honed, and applied it will produce an excellent surface finish. The high-speed steel, broad-nose tool has a deep hook ground on the face, which, in effect, gives it a very large rake angle. A small radius is ground or honed at each end of the cutting edge. The cutting edge should be honed to a sharp edge and a fine finish. Cemented carbide broad-nose tools should have a back rake angle from −6 to +15 degrees, depending

Ch. 7 SINGLE-POINT CUTTING TOOLS AND PERFORMANCE 153

on the hardness of the material being cut. The tool must be positioned with the end-cutting edge parallel to the work, as shown in the illustration, in order to contact the cylindrical work surface along the tool's entire length. It is used with a very light depth of cut which varies from approximately .001 to .005 inch. The recommended feed rate is .015 to .750 inch per revolution. Sometimes, even heavier feeds are used. The lighter feeds generally result in a better surface finish, while the heavier feeds result in higher production rates. The cutting speed used with broadnose tools should generally be lower than recommended for turning tools. The high feed rates possible with this tool cause it to traverse the length of the turned surface much more rapidly than the turning tools; therefore, there is less wear on the cutting edge which results in greater accuracy when turning long, cylindrical surfaces. The short cutting time, and the greater accuracy obtained, make the broad-nose tool very useful when turning long, cylindrical or tapered surfaces.

Grinding Single-Point Cutting Tools

Single-point cutting tools are generally ground by hand on pedestal or bench type grinders. Some of these are specifically designed to grind single-point cutting tools and have a table on which the tools can be placed while they are held against the grinding wheel by hand. The grinding wheel used on these machines is designed for grinding on the side of the wheel in order to produce a true plane surface on the workpiece. The table can be set at an angle, with respect to the side of the wheel, thereby enabling the desired angle to be ground on the tool. Provisions are usually made to have a flow of coolant available in order to prevent the tool from overheating while grinding. Most single-point, cemented-carbide cutting tools are ground on this type of grinder using a diamond-impregnated grinding wheel. High-speed-steel cutting tools can also be ground on these machines using an aluminum oxide grinding wheel.

Most high-speed-steel cutting tools, however, are ground on a simple pedestal or bench grinder. The tool bit is held by hand against the peripheral face of the wheel, at the angle required. This method is called off-hand grinding and it requires a considerable amount of skill. This skill can be acquired by analyzing what is to be accomplished and by practice.

A grinding wheel suitable for grinding high-speed steel should be mounted on the pedestal grinder. For this type of grinding a typical wheel specification is A-36-O-5-V for rough grinding, and A-60-M-5-V for finish grinding. (See Chapter 14, Volume 2, for the meaning of grinding-wheel specifications.) The grinding wheel should be dressed so that it is straight and that it runs true. Occasionally, it must be redressed in order to keep it sharp and true. A dull grinding wheel will cut more slowly and can cause the tool to become overheated. It is desirable to use a grinder equipped to supply a flow of coolant over the face of the wheel. If a coolant is not available on the grinder, the tool should be dipped in water, occasionally, to cool it. It should never be allowed to overheat to the ex-

tent of becoming discolored as this can cause metallurgical damage which will reduce the life of the tool. The most effective way of preventing the tool from becoming overheated is to control the pressure used to hold the tool against the face of the grinding wheel. A very heavy pressure can cause the tool to overheat even if a coolant is used on the machine and the grinding wheel is sharp.

The procedure for grinding a general-purpose turning tool (see view A, Fig. 7-9) by the offhand method on a pedestal grinder, will be described. The tool angles are shown in Fig. 7-12, and the grinding operation is illustrated in the series of views in Fig. 7-13.*

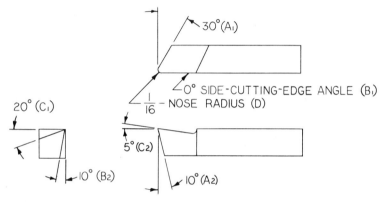

Fig. 7-12. General purpose turning tool.

Step 1. *Objective:* Simultaneously grind the 30° end-cutting edge angle (view A_1) and the 10° end-relief angle (view A_2) shown in Fig. 7-12.

Procedure: The operation is illustrated in views A_1 and A_2 in Fig. 7-13, which shows the operation from the top and side, respectively. The tool bit is firmly held in the fingers in such a position that the required angles will be ground and that the grinding wheel will cut *toward* the end-cutting edge rather than *away* from it. The fingers are allowed to rest on the toolrest of the grinder in order to steady the tool. First rough grind the tool by moving it slowly back and forth across the face of the wheel until the ground surface is within about 1/16 inch from the top of the tool.. It is then finish ground by holding the tool steady against the face of a finishing wheel, without movement, until the ground surface on the end reaches the top of the tool. The finished surface should be a single "plane." Of course, this surface will

*The views in Fig. 7-13 were obtained through the kind cooperation of Mr. Harold D. Manning, Industrial Arts Department Chairman, and Mr. Paul Vogel, student at the James E. Sperry Senior High School in Henrietta, New York.

Ch. 7 SINGLE-POINT CUTTING TOOLS AND PERFORMANCE 155

Courtesy of Mr. Harold D. Manning and Mr. Paul Vogel

Fig. 7-13. Sequence for grinding general purpose turning tool. (Upper left) Top view of simultaneously grinding end-cutting-edge angle and end-relief angle. (Upper right) Side view of same operation. (Center left) Top view of simultaneously grinding side-relief angle and side-cutting-edge angle (0°). (Center right) Side view of same operation. (Lower left) Simultaneously grinding side-rake angle and back-rake angle. (Lower right) Grinding the nose radius.

156 SINGLE-POINT CUTTING TOOLS AND PERFORMANCE Ch. 7

not be a true plane since it will conform to the curvature of the grinding wheel. Do not allow the tool to become overheated at any time.

Step 2. *Objective:* Grind the 10° side-relief angle (view B_2). (See Fig. 7-12.) The side-cutting-edge angle (view B_1), which, in this case, is 0°, is also ground simultaneously with the side-relief angle in this step.

Procedure: This operation is shown in views B_1 and B_2 in Fig. 7-13, which show the top and side, respectively. Hold the tool against the wheel so that the correct angle will be ground and so that the grinding wheel will cut *toward* the side-cutting edge rather than *away* from it. The tool is held steadily against the wheel, without movement, for both rough and finish grinding. If the side-cutting-edge angle is larger than 0 degrees, a slow back-and-forth movement across the face of the tool may be used for rough grinding. Rough grind until the ground surface is approximately $1/16$ inch from the top of the tool and then carefully finish grind until the ground surface reaches the top of the tool. The finished surface should be a single "plane" as before. Do not allow the tool to overheat.

Step 3. *Objective:* Simultaneously grind the 20° side-rake angle (view C_1) and the 5° back-rake angle (view C_2) on the face of the tool, as shown in Fig. 7-12.

Procedure: Hold the tool without movement against the wheel as shown in view C, Fig. 7-13. The angle at which the tool is held should be such as to produce the correct side-rake and back-rake angles. The side-cutting edge should be on top in order to make the grinding wheel cut toward this edge. Rough grind until the ground surface on the face is within about $1/16$ inch of the side-cutting edge. Then, very carefully, finish grind the face until a sharp edge is produced on the side-cutting edge, exercising great care not to overheat the tool. The face should be a continuous "plane" subject only to the curvature of the grinding wheel.

Step 4. *Objective:* Grind the nose radius.

Procedure: Holding the tool firmly in the fingers, place the side-cutting-edge flank gently against the grinding wheel. Then, using only a light pressure against the wheel, swing the tool around until the face of the wheel is against the end-cutting-edge flank. Repeat this movement, back and forth, until the desired nose radius is ground. (See view D in Fig. 7-13.) The entire surface below the nose must be ground to provide a relief angle for the nose. The radius should make a smooth blend with the side-cutting edge and the end-cutting edge, and with the two flanks below these cutting edges.

Ch. 7 SINGLE-POINT CUTTING TOOLS AND PERFORMANCE 157

Step 5. *Objective:* Inspect and finish dress the tool.

Procedure: The tool should be inspected visually to ascertain the condition of the ground surfaces. The cutting edge should be inspected for sharpness and for nicks or burrs. In many instances a careful visual inspection is all that is required, especially if it is performed by an experienced craftsman. At other times the angles must be measured using a precision steel protractor or a special gage made for this purpose. The nose radius can be checked by means of a radius gage.

Any imperfections found are corrected by regrinding the affected surfaces using the procedures described in the previous steps. Sometimes the face and flanks are smoothed by a few light strokes with a honing stone, being careful not to blunt the cutting edges. When the tool is to be used to take very heavy roughing cuts, one, or at the most, two light strokes can be taken across the cutting edge at a flat angle. In severe cases this will help the tool; however, excessive honing should be avoided as this will reduce the life of the tool.

Frequently, high-speed-steel, single-point cutting tools are ground by first rough grinding all of the surfaces except the nose radius. Each surface is rough ground in the same sequence as given in the steps above. The tool is then finish ground again using the sequence given in the steps, with the nose radius ground last of all.

Some tool holders, such as those shown in view A in Fig. 7-5, hold the tool at an angle instead of horizontally. In this case the end-relief angle ground on the tool must be increased so that it will be 10 degrees when the tool is held in the holder. Usually, the back-rake angle is made zero degree on cutting tools held by these tool holders; the back-rake angle being provided by the holder.

Chip Formation

The three basic types of chips formed in metal cutting are the continuous chip, the discontinuous chip, and the segmental chip, Fig. 7-14. When the cutting tool first makes contact with the metal, it compresses the metal ahead of the cutting edge. As the tool advances, the metal ahead of the cutting edge is stressed to the point where it will shear internally, causing the grains of the metal to deform and to flow plastically along a plane called the shear plane, or roughly perpendicular to the face of the tool.* When the type of metal that is being cut is ductile, such as steel, the chip will come off in a continuous ribbon, as at A, Fig. 7-14. The internal shearing action of this type of chip, called a continuous chip, causes plastic flow but it does not cause the metal to separate, or fracture;

* The shear plane is not a true plane inasmuch as it usually is slightly curved. Furthermore, it is not a surface, but a region within the metal along which internal shear occurs.

158 SINGLE-POINT CUTTING TOOLS AND PERFORMANCE Ch. 7

Courtesy of Cincinnati Milacron

Fig. 7-14. A (Upper left) Continuous chip. B (Upper right) Discontinuous chip. C (Lower left) Segmental chip. D (Lower right) Continuous chip with built-up edge.

the chip escapes by flowing over the face of the tool. The movement of the chip over the tool face, under high pressure, does further work to that portion of the chip in contact with the face. This surface, located on the underside, of the chip, is work hardened and will have a shiny appearance.

If the metal being cut is brittle, such as gray cast-iron, the stresses on the shear plane will eventually, cause the metal to fracture because it is

Ch. 7 SINGLE-POINT CUTTING TOOLS AND PERFORMANCE 159

unable to flow plastically as can a more ductile material. This causes the chip to break up into particles as shown at B, Fig. 7-14. Such a chip is called a discontinuous chip. The formation of a discontinuous chip is illustrated in Fig. 7-15.

Courtesy of Cincinnati Milacron

Fig. 7-15. The formation of a discontinuous chip. (Left) Cutting tool with a large rake angle. (Right) Cutting tool with a small rake angle.

A segmental chip is frequently encountered when machining ductile materials. It is shown at C, Fig. 7-14. When a heavy feed is used, the stresses on the shear plane cause the metal in this plane to eventually fracture internally. This may be further abetted either by low frequency and high amplitude vibrations of the cutting tool or the work, or both. The internal fractures will spread to the outside surfaces; however, they are also quickly welded together again by the action of the pressures existing in this region. The result is a continuous chip that has a file-tooth appearance on one side. The side of the chip that passes over the tool face is more completely welded together, and is burnished by the face so that it has a smooth appearance. The degree and depth of the segmentation of a chip can vary greatly because of the feed, the cutting speed, and the nature of the metal being cut.

Another phenomenon associated with the formation of the chip is the built-up edge shown at D, Fig. 7-14. This generally occurs when cutting ductile materials at relatively low (approximately, 150 to 200 fpm, or less) cutting speeds. The built-up edge is a highly work-hardened layer of the work material that has been welded onto the cutting edge. It starts out as a small layer that gradually builds up in size by having more work material continuously added to it by pressure welding. Eventually, the built-up edge enlarges to the point where it becomes unstable and it breaks off. When this happens, it may pass off with the chip or it may wedge between the work and the cutting edge to score the work surface. It can also pluck out small particles of the cutting tool material as it breaks off, resulting in a deterioration, or wear, of the cutting edge. When the built-up edge has broken off, a new built-up edge starts to form and this cycle is repeated. In some cases, the size of the built-up edge becomes quite large and protrudes beyond the cutting edge to score the surface of the work while it is still attached to the tool.

The effect of the built-up edge is to cause the machined surface of the work to be rough. *The built-up edge is the principal cause of surface roughness on the machined surface.* Efforts to achieve a good surface finish should be directed toward eliminating the formation of the built-up edge. It will *not* form when the cutting speed exceeds a certain critical rate which can be easily reached with cemented-carbide cutting tools. When these tools are operated above this speed, an excellent surface finish is obtained without difficulty. As high-speed steel must usually operate below this cutting speed, in order to obtain a good surface finish, a cutting fluid is used with high-speed steel tools, to reduce or to eliminate the built-up edge. Using a large rake angle, a light feed rate, and having a smooth surface finish on the cutting edge will also help. Since the formation of the built-up edge requires a metallurgical interaction between the work material and the cutting tool material, using a cutting tool material that does not react with the work material will also reduce or eliminate the built-up edge. Industrial diamonds fall into this category; they will produce excellent surface finishes; however, their use is restricted to very light finishing cuts.

Chip Control

Chip control is not a problem when cutting metals that form a discontinuous chip. It is only troublesome when machining metals that form a continuous or a segmental chip with high-speed-steel cutting tools. Chip control, on the other hand, is essential when forming continuous or segmental chips with cemented-carbide cutting tools. The high cutting speeds used cause such a rapid formation of the chip that both the machine operator and the equipment are endangered.

The action of a plate-type chip breaker is shown in Fig. 7-16. The plate deflects the chip, causing it to curl, and to strike against the flank

Courtesy of Kennametal Inc.

Fig. 7-16. Chip control with plate type chip breaker on throwaway insert tool, when cutting steel. (Left) .014 ipr feed. (Right) .036 ipr feed.

of the tool holder or the chip breaker. When the chip strikes the tool holder it breaks off into short lengths which are curled and have an appearance like a figure 9. The size of the broken off chips depends on the feed rate and the design of the chip breaker. Chip breakers may be ground into high speed steel and into cemented-carbide tools to cause the chip to curl and break off as described. The three basic types of chip breakers are shown in Fig. 7-17.

Tool Wear

All metal-cutting tools begin to wear almost at the instant that they begin to be used. Tool wear cannot be prevented when cutting metal. When the tool wear is kept within acceptable limits, the cutting tool can continue in service. Usually, the initial amount of tool wear occurring within the first minute of cutting is quite large. The wear then continues to progress at a slower but uniform rate. The wear rate for a given tool material and work material at this stage is primarily dependent upon the cutting speed. Higher cutting speeds result in faster tool-wear rates

162 SINGLE-POINT CUTTING TOOLS AND PERFORMANCE Ch. 7

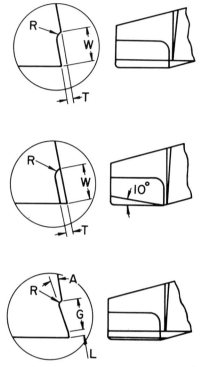

Courtesy of Kennametal Inc.

Fig. 7-17. Basic types of chip breakers. (Top) Parallel type. (Center) Angular type. (Bottom) Groove type.

and if the cutting speed is too high, the wear rate will become excessive, and the tool life will be short. When the wear on the flank of the cutting tool becomes great, it will burnish the surface of the work causing it to work-harden. This is sometimes quite noticeable by a glazed or shiny appearance. When the wear on the flank of the tool becomes excessive, it will rub on the work surface, thereby causing the temperature of the tool to increase rapidly. The rapid increase in temperature causes the tool material to soften and quickly fail in a so-called catastrophic tool failure. The cutting tool should be replaced before such a catastrophic failure occurs. The amount of tool wear can be conveniently measured on the flank of the tool. Recommendations for the amount of flank wear that determines the end of tool life are given in the "U.S.A. Standard Specifications For Tool Life Testing With Single-Point Tools" (USAS B5.19-1946 and B5.34-1956) published by The American Society of Mechanical Engineers. For finish-turning tools the amount of flank wear should generally not exceed .006 to .014 inch; and for rough turning tools, the amount of flank wear should not exceed .020 to .040 inch.

Ch. 7 SINGLE-POINT CUTTING TOOLS AND PERFORMANCE 163

More specific recommendations are available in the standard mentioned above.

The fundamental causes of tool wear are due to very complex mechanical, chemical, and metallurgical phenomena. These need not concern the machinist or the toolmaker. He should, however, be able to recognize the visible manifestations of tool wear.

Flank Wear. This is wear always occurring on the flank of the cutting tool when cutting metal. It can readily be measured with a low-powered microscope (15 to 30×) having a suitable linear, or micrometer scale. As already explained, the rate at which the size of the flank wear progresses is uniform, within broad limits, except when an excessive cutting speed is used. The time when a cutting tool should be replaced can be determined by the amount of the flank wear; which is the measured distance below the cutting edge to which it extends. The flank wear rubs against the surface of the work causing it to be burnished and to generate heat, which is transferred to the tool and to the work. When the flank wear is excessive, this heat will cause the catastrophic failure of the tool. When cutting within acceptable limits of flank wear, however, the heat generated by this rubbing action is small, compared to the heat that is generated on the shear plane where the grains of metal in the work are severely deformed in order to form the chip.

Crater Wear. This is the formation of a rather large crater on the face of the tool as shown in A, Fig. 7-18. The crater is formed by the combined effect of heat, abrasion caused by the chip sliding over the face of the tool, and some very complex metallurgical interactions between the work material in the chip and the tool material itself. The severity of cratering is dependent upon the cutting tool material used, the work material, and the cutting speed used. The crater always forms at a very short distance behind the cutting edge. It increases in size as the tool continues to cut and grows deeper and larger in width. The depth of the crater is usually difficult to determine because work material from the chips is welded into the crater and partially or almost completely fills it up. Ultimately, the crater will weaken the side-cutting edge and cause parts of it to break off. This is a prelude to complete tool failure. Very often, before the crater causes a failure of the side-cutting edge, it will grow in the direction of the end-cutting edge resulting in a breakthrough in this region. (See B, Fig. 7-18.) Crater breakthrough at the end-cutting edge is followed by a rapid failure of the nose of the cutting tool and catastrophic tool failure if the tool is allowed to continue in use. It should be replaced preferably before crater breakthrough occurs and most certainly after this has happened. Crater breakthrough can be delayed by simply using a relatively small end-cutting edge angle on the cutting tool.

Pullouts. Pullouts, shown at D, in Fig. 7-18, are small particles of work material that are plucked from the face of the tool, leaving a very small

164 SINGLE-POINT CUTTING TOOLS AND PERFORMANCE Ch. 7

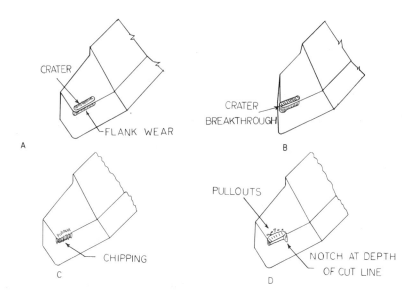

Fig. 7-18. Visible manifestations of tool wear. A. Flank wear and crater wear. B. Crater break-through at end-cutting edge. C. Cutting-edge chipping. D. Notching at depth of cut line and pull-outs on face of tool. Note welded-on layer of material that is not a built-up edge.

crater. These craters are much smaller than the crater associated with the cratering phenomenon described above. The small craters caused by the pullouts can occur within the larger crater, although they also occur in other areas of the tool face.

Pullouts are caused by particles of work material that are firmly welded to the face of the tool. This can occur as part of the built-up edge, in the crater, and also in other areas which perhaps only occasionally come in contact with the chip. These particles are sometimes broken off again by the chip, by impact, or possibly by rewelding to the chip. When this occurs, the particle of work material remains firmly attached to the tool material causing the tool material to be plucked out leaving a small hole or crater. Sometimes these small craters are visible, while at other times, they rapidly fill with work material which is deposited by the chips and welds to the cutting tool material within the crater. In either case, however, there is deterioration and damage to the surface of the cutting tool.

Cutting Edge Chipping. This is evidenced by small chips breaking off the cutting edge (see C, in Fig. 7-14). It is caused by the mechanical failure of the cutting edge to withstand the load to which it is subjected when taking a cut, or by small parts of the cutting edge being broken off with the built-up edge. The areas of the cutting edge that are chipped,

Ch. 7 SINGLE-POINT CUTTING TOOLS AND PERFORMANCE

usually are at least partially filled with work material that is welded in place. The mechanical failure of the cutting edge can be due to any one or a combination of causes: by using a rake angle or a relief angle that is too large, and using a very hard and brittle cutting-tool material. When chipping occurs, the rake and relief angles must be reduced, especially when very hard and brittle tool materials are used. Very hard tool materials may also require that the feed rate be reduced in order to decrease the load on the cutting edge.

It is sometimes necessary to hone a very small (.002 to .004 inch wide) bevel on the cutting edge. This strengthens the cutting edge; however, it will also reduce the tool life obtained for the cut. For this reason the cutting edge should never be honed to a bevel or a radius unless the severity of the cut demands that the cutting edge be strengthened to prevent very early tool failure due to chipping. This frequently must be done when cutting very hard work materials and to resist impact loads when taking heavy, interrupted cuts.

Other forms of tool failure that occur, but are not too frequently encountered in turning, are thermal cracking, notching at the depth of cut line, and compression of the cutting edge. Thermal cracking is caused by rapid fluctuations in the temperature of the cutting tool which result when the cutting edge penetrates and leaves the workpiece in rapid succession. This action is encountered when milling, and in turning when taking interrupted cuts. Small cracks form, a short distance away from the cutting edge, and propagate toward the cutting edge causing it to fail. Thermal cracks can also be caused by improper tool grinding. Notching at the depth of the cut line is the formation of a significantly larger flank-wear land on the edge of the tool that is in contact with the original surface of the workpiece. (Shown at D, in Fig. 7-18.) It can occur when the tool is cutting through a layer of highly work-hardened material or when cutting through a layer of hard, abrasive material such as scale, found on castings and hot-rolled bars. Notching at the depth of cut line also occurs when turning certain high-temperature alloys. In this case, it can be prevented by using a large lead angle. Compression of the cutting edge occurs when taking very heavy cuts with carbide cutting tools. The heat and pressure generated causes the carbide in the region of the cutting edge to be compressed below its original height.

QUESTIONS

7-1. What is meant by the term "tool life"?

7-2. Name eight factors that affect the cutting speed with which a cutting tool can cut.

7-3. What work-material factors affect the cutting?

7-4. Why can most work materials cut successfully using a fairly wide range of cutting speeds?

7-5. What is the effect of cutting fluids when cutting metal?

166 SINGLE-POINT CUTTING TOOLS AND PERFORMANCE Ch. 7

7-6. What effect do the feed and the depth of cut have on the recommended cutting speed, when cutting a given metal?

7-7. What is the effect of feed, depth of cut, and cutting speed on tool life?

7-8. Name six basic types of cutting-tool materials.

7-9. What distinguishes a crater resistant, tungsten-carbide, cutting-tool material from other tungsten carbides?

7-10. Name the type of cemented carbide that you would use to cut the following metals:
 a. Gray cast-iron
 b. Ferritic malleable iron
 c. Pearlitic malleable iron
 d. AISI 1020 steel
 e. AISI 4140 steel
 f. Austenitic stainless steel
 g. Annealed, high-speed tool steel
 h. Martensitic stainless steel
 i. Aluminum
 j. Brass

7-11. What is the advantage of using diamond, single-point cutting tools?

7-12. Name the six principal angles of a single-point cutting tool.

7-13. What distinguishes a right-hand from a left-hand single-point cutting tool?

7-14. Why are the side relief angle and the end relief angle, ground on single-point tools?

7-15. What is the best average value of the relief angle for:
 a. high-speed cutting tools?
 b. cemented-carbide cutting tools?

7-16. What is the difference between positive- and negative-rake angles?

7-17. What is the advantage of having a positive side-rake angle?

7-18. Why are negative-rake angles used?

7-19. What is the difference between the side-cutting-edge angle and the lead angle?

7-20. Describe the effect of the lead angle on the performance of single-point cutting tools.

7-21. Name the causes of chatter.

7-22. How can chatter be eliminated?

7-23. What is the effect of the nose radius on the performance of single-point cutting tools?

7-24. Why is the cutting tool sometimes positioned for turning, with the nose leading?

7-25. How is the combination roughing-finishing tool used?

7-26. How is the broad-nose finishing tool used?

7-27. Name the three basic types of chips that are formed when cutting metal.

7-28. How is the chip formed when cutting metal?

7-29. What is the built-up edge?

Ch. 7 SINGLE-POINT CUTTING TOOLS AND PERFORMANCE 167

7-30. What is the principal cause of roughness on a finish-turned surface?

7-31. How can the surface finish be improved?

7-32. What is meant by flank wear?

7-33. What is the difference between crater wear and pullouts?

7-34. How does the crater breakthrough at the end-cutting edge occur?

7-35. What is the cause of cutting-edge chipping?

7-36. What is meant by notching at the depth of cut line?

7-37. Calculate the lathe-spindle speed for cutting an ASTM Class 40 gray cast-iron using a cemented-carbide cutting tool. The diameter of the work is 3 inches, the feed is .031 ipr, and the depth of cut is .250 inch.

7-38. Calculate the spindle speed for cutting H-10 hot-work tool steel with a high-speed cutting tool using a feed of .015 ipr and a depth of cut of .100 inch. The diameter of the workpiece is .750 inch.

7-39. A 6-inch diameter part made from AISI 4140 steel having a hardness of 220 Bhn is being turned in a lathe using a cemented-carbide cutting tool. The spindle speed used is 300 rpm, the feed is .020 ipr, and the depth of cut is .150 inch. The tool life obtained is very short. Determine the cause, and recommend a new cutting speed.

CHAPTER 8

Cylindrical Turning

The term "cylindrical turning" is used to describe a variety of machining operations that are performed by rotating the workpiece which is held in a lathe. These operations may be performed on either the outside or the inside of the workpiece, as shown in Fig. 8-1. The outside diameter, or O.D., operations are shown at A, and the inside diameter, or I.D., operations are shown at B. Not all of these operations will be treated in this chapter; however, they will all be covered in subsequent chapters. This chapter will concern itself with the basic outside-diameter operations.

Turning operations can be classified as rough turning, or as finish turning, operations. The objective of the rough turning operation is to remove the bulk of the excess metal as rapidly and as efficiently as possible. Enough metal must be left on the workpiece so that a finish turning operation can be performed. The objective of the finish turning operation is to attain the required dimensional tolerances and the required surface finish on all surfaces of the workpiece. Parts are sometimes semi-finish turned in a lathe, and the final dimensions and surface finish are obtained on a cylindrical grinding machine or on an internal grinding machine. Precision work can, however, be done on a lathe when the basic principles of lathe operation are understood and when this work is performed with skill.

Preparation of the Workpiece

The workpiece may be held in the lathe in several different ways. It can be clamped in a chuck, bolted to a faceplate, held between two centers, or held in a chuck or faceplate at one end and supported by a tailstock center at the other end. Additional work support may be obtained by a follower rest or by a steady rest. Complicated parts, that would otherwise be difficult to hold, are sometimes held in special lathe fixtures. This chapter is concerned, primarily, with the operation of the lathe when the work is held on centers. Long, slender workpieces must always be well supported when held in a lathe. They may be held between two centers, one at the headstock and one at the tailstock. Larger parts, such as are shown in Fig. 8-2, are often held in a chuck or by the faceplate chuck jaws at the headstock end instead of a center, because this method of holding is more rigid and also because very large lathe dogs are often difficult to secure to the workpiece. The tailstock end of this workpiece is held by a

Ch. 8 CYLINDRICAL TURNING 169

Fig. 8-1. A (Top) The basic outside-diameter turning operations. B (Bottom) The basic hole producing lathe operations.

tailstock center. Smaller workpieces such as shown at B, in Fig. 8-9, are often held between two centers using a lathe dog which is clamped to the work and driven by the driver plate.

Before the workpiece can be placed between centers in a lathe a center hole must be drilled in each end. This hole is drilled with a combination drill and countersink which is commonly called a center drill (see Fig. 8-3). The conical hole should be drilled deep enough so that a good bearing surface is provided for the lathe centers; it must not, however, be

Courtesy of The R. K. LeBlond Machine Tool Company

Fig. 8-2. Measuring the diameter of a workpiece in a lathe. The part is held by faceplate jaws at the headstock end.

drilled beyond the largest diameter of the center drill. The countersink has a 60-degree included angle and the lead drill provides a clearance for the lathe center points. The axes of the two center holes must coincide with each other so that the lathe centers will seat firmly on the conical surfaces of the holes. There are several methods of drilling the center holes in the work. Three different methods for doing this job will be described.

Fig. 8-3. Combination drill and countersink, or center drill, and its application in centering parts for turning.

1. The centers of the end faces of the work are determined by laying out, as described in Chapter 3. A deep, center punch-mark is made at the center of each end. A drill chuck is mounted in the headstock spindle of the lathe as shown in Fig. 8-4. The center punched hole of the workpiece is held against the tailstock center, as shown, and the tailstock spindle is advanced in order to feed the other end face into the center drill. One hand is used to turn the tailstock handwheel while the other hand is used to hold the work. The center drill is started in the center punch mark and the hole is drilled

Fig. 8-4. Drilling the center holes in the end of a cylindrical workpiece.

to the correct depth. The work is then turned around and the other center hole is drilled. This is an accurate method of drilling the center hole when the workpiece is light enough in weight to be easily handled.

2. Short workpieces that can be held in a chuck without protruding too far, can be center drilled, as shown in Fig. 8-5. The workpiece must be accurately held in the chuck, otherwise the axes of the two center holes will not coincide. If the three-jaw universal chuck does not run true, it is best to use a four-jaw, independent chuck that can be adjusted to hold the workpiece concentric within one-thousandth of an inch, if desired. This is accomplished by placing a dial test indicator against the workpiece and adjusting the chuck until the hand of the indicator does not move when the lathe spindle is slowly turned by hand. A center drill is placed in a drill chuck which is held in the tailstock spindle. The tailstock spindle is advanced manually to drill the hole to the correct depth.

Courtesy of The R. K. LeBlond Machine Tool Company

Fig. 8-5. Drilling the center holes in the end of short workpieces by holding the workpiece in a chuck.

This method has the advantage of not requiring a layout to be made on the workpiece prior to drilling.

3. Long workpieces, and those that are too heavy to be held by hand, are center drilled as shown in Fig. 8-6. The work is held by a three-jaw, universal chuck or by a four-jaw, independent chuck. It must be held concentric with the rotation of the lathe. The other end of the workpiece is supported by a steady rest. The steady rest must be adjusted to support the work so that its axis coincides with the axis of the lathe. One method by which this can be done is to find the center of the end face of the work by making a layout as described in Chapter 3. A tailstock center is then positioned opposite the layout mark which determines this center and the steady rest is adjusted until this layout mark is in alignment with the point of the tailstock center. The work can also be aligned by using a dial test indicator that is attached to a surface gage. The contact point of the indicator is moved over the top of the work, and along the side of the work, at each end. When indicating the side of the work, parallelism between the work and the axis of the lathe is obtained by using the finish machined vertical sides of the lathe bed as a reference surface for the surface gage. Accurate results can be obtained by using a surface gage

Ch. 8 CYLINDRICAL TURNING 173

without an indicator. The point of the surface gage scriber is adjusted to touch the top and side of the work at each end. This method is preferred when the work has a rough surface scale.

Before the center holes are drilled, a facing cut should be taken across the end of the work. This is done as shown at I, Fig. 7-9. The work can be machined to the correct length by facing both ends. A short spot should be turned on the outside diameter adjacent to the end face as shown in Fig. 8-6. The spot is turned by cutting just deeply enough to produce a cylindrical surface that is concentric with the center hole in the end. The purpose of the spot is to provide a reference surface which can be gripped by the chuck and against which an indicator can be placed in order to test the concentricity with which that end of the work is held by the chuck. The center hole is drilled in each end using the tailstock, as shown in Fig. 8-6.

Preparation of the Engine Lathe

The engine lathe must be level in two directions in order to produce accurate work. If it is not level the bed will have a twist which will cause the headstock and the tailstock to be misaligned with the movement of the carriage and the cross slide. A precision level that is accurate to .0005 inch per foot is used to level the lathe. It must be level, parallel and perpendicular to the lathe axis at both extremes of the lathe bed. Level-

Courtesy of The R. K. LeBlond Machine Tool Company

Fig. 8-6. Drilling the center holes in the end of larger workpieces.

ing screws or wedges are provided on the legs of the lathe to level the bed. The lathe should be leveled when installed, and a periodic check should be made to make certain that it remains level. It is not, however, necessary to check this before doing every job.

If the workpiece is to be held between the headstock and the tailstock center the following preparations must be made. The taper in the headstock must be thoroughly cleaned before the center adapter is installed. The cleanliness of the headstock taper, as well as the taper of the adapter should be checked by using the bare hand or fingers. The presence of nicks or small particles of dirt or metal could not be detected through a rag or cloth placed on the tapers; however, they can readily be felt by the fingers or the bare hand. All taper fits on any machine tool should be checked in this manner. If the taper is dirty, or badly nicked, it will not hold and it will not provide the desired accurate alignment of the mating parts. After the adapter has been inserted in the lathe spindle, the headstock center is placed in the adapter using the same care.

The headstock center of the lathe always rotates with the lathe spindle; therefore it is called a live center. Tailstock centers may be of the type that also rotate with the workpiece; however, frequently they are stationary or dead centers. The cone point of the headstock center must rotate without any eccentricity otherwise the two ends of the finish-turned workpiece will not be concentric to each other. Such eccentricity cannot be tolerated on precision parts.

To show how this eccentricity can occur, assume that the workpiece is held in a lathe between a live headstock center and a dead tailstock center. Furthermore, assume that the tailstock center is accurately aligned with the axis of the lathe but that the headstock center is not rotating concentrically with respect to the lathe axis. The longitudinal feed, guided by the ways of the lathe, will cause the path of the cutting tool to be parallel to the lathe axis. As this tool cuts the rotating workpiece it will produce a cylindrical surface which is concentric to the tailstock center and to the center hole in the tailstock end of the workpiece. When the workpiece is turned end for end in the lathe, the center hole, which is concentric with the first-turned surface, will now be mounted on the eccentrically rotating headstock center which will cause the first turned surface to rotate eccentrically. While the first-turned surface is turning eccentrically, the cutting tool will turn the other end of the workpiece concentric with the axis of the lathe as before. Thus, the surfaces turned at each end of the work cannot be concentric to each other. This can only be corrected by truing the headstock center so that it rotates concentrically, with respect to the lathe axis.

Headstock centers may be made from soft steel or from hardened steel. Soft centers have the advantage of being easily trued by simply turning a 60 degree cone point, while mounted in the headstock of the lathe as shown at A in Fig. 8-7. The compound rest is positioned at 30 degrees, with respect to the lathe axis, and a cut is taken by feeding the tool with the compound rest slide. Only very shallow cuts should be taken until the

Fig. 8-7. A. Truing a soft headstock center with a lathe tool. B. Truing a hardened headstock center by grinding with a tool-post grinder.

cone point of the center runs true. With the compound rest in this position, the cone point of the center will have an included angle of 60 degrees. This angle must be accurate and should be checked with a center gage or with a precision bevel protractor.

Hardened headstock centers usually retain their accuracy for a longer period. They are trued in a manner that is similar to truing a soft center except that the cone point is ground with a grinding wheel instead of a single-point tool as shown at B, in Fig. 8-7. The grinding wheel is driven by a tool post grinding attachment which is mounted on the compound rest. The depth of cut of the grinding wheel is adjusted with the cross feed and the wheel is fed across the cone point of the center by the compound rest slide. The cone point is ground until it runs true and the included angle is correct. Hardened tailstock centers can be ground in the same manner by mounting them in the headstock spindle of the lathe.

The tailstock may be a live center that rotates with the workpiece or it may be a dead center that does not rotate. The live tailstock center, shown in Fig. 8-8, is mounted in precision anti-friction bearings. Live tailstock centers are a necessity when turning at very high cutting speeds such as those normally used with cemented carbide cutting tools. The cone point of the live tailstock center must run true for the same reasons that the headstock center must run true. This is checked with a dial test indicator. The center is slowly turned, manually, and any movement of the indicator hand is noted. If the center does not run true it must be replaced. It must be disassembled and reground in order to be trued. Dead tailstock centers are always hardened because the cone point acts as a bearing surface for the rotating workpiece. All the surfaces of the dead center must be smooth; they are finished by grinding.

Another fundamental requirement for turning a cylindrical workpiece that is held on the tailstock center is that this center must be aligned to

Courtesy of the Ideal Industries Inc.

Fig. 8-8. Live tailstock center mounted in precision antifriction bearings.

coincide with the axis of the lathe. If the work is also held by a center at the headstock end, the headstock and the tailstock centers must be aligned with each other. If the tailstock center is misaligned, the turned surface on the work will be tapered instead of cylindrical. For every .001 inch that the tailstock center is misaligned in a direction crosswise to the lathe bed, the workpiece will be tapered .002 inch in its length. It can thus be seen that the tailstock center must be very accurately aligned in a crosswise direction. An error in the vertical alignment of the tailstock center is not nearly as critical when turning. For example, to cause a taper of .001 inch to be turned in the length of a workpiece that is one inch in diameter, the vertical misalignment must be .022 inch; for a 4-inch diameter, the vertical misalignment must be .045 inch. The vertical alignment built into precision lathes is within a few thousandths of an inch and the vertical misalignment of the tailstock center is almost never a cause of taper unless the lathe is very badly worn. However, a small vertical misalignment can cause a straight reamer, which is held in the tailstock spindle, to ream a slightly tapered hole.

A check should be made to determine the accuracy of the alignment of the tailstock center before any attempt is made to turn a cylindrical surface unless it is known to be in alignment. If the tailstock center is not in alignment, steps must be taken to bring it into alignment. There are two methods of aligning the tailstock center; these are described later in this section. First, however, as a rough check, the tailstock can be moved up to the headstock until the two centers are opposite each other and their alignment can be checked visually. If a headstock center is not being used, another method of making a rough check is to check the alignment scale at the end of the tailstock (see Fig. 6-16). If a misalignment is indicated, the clamp bolt should be loosened, and the two adjusting screws turned, until the proper alignment is obtained.

A more precise check for the alignment of the tailstock should be made before attempting to turn an accurate cylindrical surface. The recom-

mended procedure for *checking this alignment* is to cut two spots on the workpiece as shown at A and B in Fig. 8-9. The two spots should be cut as far apart as possible on the workpiece. In the case of a piece of cylindrical bar stock, as shown in Fig. 8-9, one of the spots should be cut as close to the lathe dog as possible while the other spot is cut at the tailstock end. If a casting or forging is being turned, the spots should be cut at each end of the longest cylindrical surface to be turned. The procedure for making a precision check of the alignment is detailed step-by-step below:

1. Set up the lathe and the cutting tool as shown at A and B, Fig. 8-9. Check to make sure that the lathe dog will not hit the carriage when the cutting tool is close to the dog. Also, check to make sure that the cutting tool can be positioned clear of the tailstock end of the workpiece in order to be able to cut the spot at that end. Adjust the cutting speed and the feed of the lathe. Use a feed of approximately 0.005 inch per revolution.
2. Position the cutting tool approximately one inch to the right of the lathe dog.
3. Start the lathe spindle and slowly feed the cutting tool into the workpiece, by hand, until a complete cylindrical surface is being turned. Then, engage the longitudinal feed and allow the tool to cut as close to the dog as possible, before disengaging the feed.

Fig. 8-9. Procedure for aligning the tailstock center: A (Upper left) First step—cutting spot at headstock end of work, B (Upper right) Second step—cutting spot at tailstock end of work, C (Lower left) Indicating headstock spot, and D (Lower right) Indicating tailstock spot.

4. Before withdrawing the tool read the setting of the micrometer cross-feed dial and note this reading. It is often convenient, when cutting the spot at the headstock end, to set the cross-feed dial to zero as this reading can be easily remembered.
5. Withdraw the cutting tool from the workpiece with the cross feed and move the carriage until the cutting tool is clear of the tailstock end of the workpiece.
6. Position the cutting tool to take the second cut by moving the cross slide until the micrometer cross-feed dial reading is the same as noted in Step 4.
7. Cut the second spot at the tailstock end using the automatic longitudinal feed. The length of this spot should be approximately 3/4 inch.
8. Measure the diameter of both spots with micrometer calipers. The centers are in alignment if the measurements on both spots are equal to each other. If the diameter of the tailstock spot is larger than the diameter of the headstock spot, the tailstock is offset crosswise *away* from the lathe operator, and to be corrected, it must be moved toward the operator. If the tailstock end is smaller in diameter than the headstock end the tailstock is positioned too far *toward* the operator, and to be corrected, it must be moved away from him. In either case, the distance that the tailstock must be moved is approximately one-half the difference in the diameters of the spots at the headstock and tailstock ends of the work.

The advantage of the procedure just outlined for checking the alignment of the tailstock center is that the time required to cut the entire length of the workpiece is saved but the effect is the same. The spot at the headstock end should always be cut first because it is easier to reproduce the depth of cut setting of the tool at the tailstock end, where the cutting tool is clear of the workpiece.

Two methods of *correcting the alignment* of the tailstock center will now be described; either one will produce quick and accurate results when skillfully carried out:

Method 1. Place a dial test indicator against the side of the tailstock spindle. The position of the indicator should be as shown in Fig. 8-10; however, the workpiece should not be removed from the lathe. The hand of the dial indicator should be set to read zero. The tailstock clamp bolt (Fig. 6-16) should then be loosened, and the tailstock adjusted crosswise by turning the two adjusting screws located on the side of the tailstock body. The amount of this adjustment should be approximately one-half of the difference in the diameters of the two spots. The dial test indicator reading is used to determine the distance that the tailstock is moved. When the clamp bolt is again tightened, the dial test indicator should read the exact distance it is desired that the tailstock be moved.

CYLINDRICAL TURNING

The entire procedure consisting of cutting a spot at the headstock and at the tailstock ends of the workpiece, measuring the difference in the diameters of the spots, and adjusting the tailstock accordingly, is repeated until the two diameters are equal.

Method 2. Two spots are cut on the workpiece as before; however, this time the diameter of the spot at the headstock end is measured with a micrometer caliper before the spot at the tailstock end is cut. The diameter of the spot at the tailstock end is then turned, to be equal to the diameter of the spot at the headstock end of the workpiece. After the two

Fig. 8-10. Dial test indicator placed against tailstock spindle when adjusting tailstock to remove taper.

spots have been turned to the same diameter, clamp a dial test indicator in the tool post of the lathe as shown at C and D, in Fig. 8-9. The indicator is placed against the first spot, as shown at C. The indicator is adjusted to read zero by moving the cross slide. The reading on the micrometer cross-feed dial is then observed and noted. On many lathes this dial may be loosened and turned to read zero. The indicator is then backed away from this spot by retracting the cross slide and the carriage is moved longitudinally until the dial indicator is opposite the second spot. The cross slide is moved toward the workpiece until the micrometer, cross-feed dial setting is the same as previously noted. In this position, D, Fig. 8-9, the dial indicator reading will show the exact number of thousandths-of-an-inch that the tailstock is misaligned, or, if the indicator readings are the same at both ends, it will prove that the tailstock is in alignment. If the tailstock is not in alignment, the dial test indicator is kept in this position and the tailstock is moved crosswise, until the hand of the dial test indicator points at zero. This test is then repeated until the reading of the dial test indicator is zero at each end of the workpiece without requiring any further crosswise adjustment of the tailstock. It is important that the zero indicator readings at each end are obtained with the tailstock clamped firmly to the bed.

Setting-Up the Work for Turning Between Centers

The correct setup for turning between centers is shown in Fig. 8-11. A driver plate is mounted on the spindle nose of the lathe. The driver plate is a small face plate that is used to drive the lathe dog.

Lathe dogs are used to transmit the drive from the driver plate of the lathe to the workpiece when turning between centers. Several different types of lathe dogs used, are shown in Fig. 8-12. The lathe dog is placed over the workpiece but is not tightened until later.

When a nonrotating, or dead, tailstock center is used, a lubricant must be used between this center and the workpiece. Although proprietary lubricants are available, a good tailstock lubricant can be made by mixing a small amount of lubricating oil with painter's white lead. No lubricant is required if a live tailstock center is used nor is a lubricant used on the headstock center. The workpiece is then placed between the centers. The lathe dog is positioned with the bent end, or tail, between one of the slots on the faceplate and is clamped to the work. The tail of the dog must not bind in the bottom of the driver plate slot, as shown in Fig. 8-13, as this will cause the headstock center to become unseated. When this occurs a larger lathe dog must be used. The tailstock spindle is adjusted until the

Courtesy of the Clausing Division of the Atlas Press Company

Fig. 8-11. The setup for turning between lathe centers.

Ch. 8 CYLINDRICAL TURNING 181

Fig. 8-12. Types of dogs, or drivers, commonly used in connection with lathe work.

work seats on both centers, without binding, and the spindle is clamped in place. It is important to check this adjustment frequently, during the course of the turning operation. When a heavy cut is taken, some of the heat generated by the cutting action is absorbed by the workpiece which

Fig. 8-13. Dog that is too short for driver plate.

will cause it to expand in length enough to make the centers bind. When this occurs, the tailstock center will overheat and, consequently, be ruined. When a heavy roughing cut is followed by a light finishing cut, the work will lose heat and contract, causing the adjustment of the centers to be too loose. A turning tool (Figs. 7-9 and 7-10) is selected and clamped in the tool post. The overhang of the cutting edge beyond the tool post should be kept to a minimum in order to make the setup as rigid as possible. The nose of the cutting tool should be positioned "on center" or at the height of the lathe centers. The turning tool usually is held about square with the work unless there is some advantage in holding it at an angle.

The Machine Setting

The engine lathe must be set to operate at the desired cutting speed and feed rate. The selection of the correct cutting speed has been treated in Chapter 7 (see Tables 7-1, 7-2, and 7-3). The spindle speed of the lathe is calculated using Formulas 7-1 and 7-2a, and the lathe spindle is set at a speed which is as near as possible to the calculated value. The feed rate to be used is governed by the requirements for surface finish on the work, the available power on the machine, the size of the cutting tool, the size of the workpiece, and by how firmly it is held in the lathe. The selection of the feed is not subject to calculation such as the spindle speed; it is determined largely by judgement based upon experience. Only the limitations imposed on the feed by the available power of the machine can be calculated. (This topic is treated later in this chapter.) The quick change, or feed-thread gear box, is then set to obtain the selected feed.

End Facing with the Work Held Between Centers

The first operation that is generally performed after the work has been set up is to face the ends of the work to length. If the ends have been faced while drilling the center holes, this operation is, of course, unnecessary. An end-facing tool, Fig. 8-14 and F, Fig. 7-9, is placed in the lathe

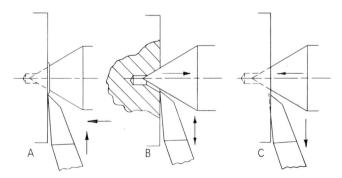

Fig. 8-14. Procedure for cutting of burr surrounding center hole, using an end-facing tool.

Ch. 8 CYLINDRICAL TURNING

tool holder. It is positioned with the nose leading slightly, as shown in Fig. 8-14. The end of the work is rough faced using the longitudinal feed handwheel on the apron to establish the depth of the cut and cross feed, in order to feed the tool across the end face. The direction of the cross feed should be from the center toward the outside diameter of the work. The rough-facing operation will leave an unsightly burr surrounding the edge of the center hole of the work, which must be removed during the finish-facing operation.

The procedure for removing this burr is illustrated in Fig. 8-14. The end-facing tool is positioned near the tailstock center, as shown at A. The tool is then moved into the face of the work using the longitudinal feed handwheel on the apron in order to obtain the depth of cut. The following steps must then be taken in rapid order:

1. The tailstock spindle is unclamped.
2. The tailstock center is moved out a very short distance by moving the spindle with the tailstock handwheel.
3. Simultaneously with Step 2, the facing tool is fed toward the center by manually turning the cross-feed handle. See B, Fig. 8-14. This procedure will cause the burr surrounding the center hole to be cut off.
4. Immediately after the burr is cut off the facing tool is backed out, again using the manual cross feed, and the center is repositioned in the center hole.
5. The finish-lacing cut is then completed by facing from the center toward the outside, as shown at C. The power cross feed may be used to take this cut.

The work is turned around in the lathe in order to face the second end. A simple layout is made before this operation is begun. A line is laid out from the finished face of the work to some distance less than the required length of the work, as shown in Fig. 8-15. It is convenient to make this distance one inch less than the required length of the work. It would be difficult to scribe a line at the exact length of the work as it would fall too close to the existing face. Furthermore, the chips resulting from the facing cut might possibly destroy the layout before the facing operation is completed. After the layout has been made, the facing operation is performed, as described above, until the distance from the second face to the layout line is one inch, as measured by a rule or by hermaphrodite calipers.

Cylindrical Turning

As already explained, two types of cuts are normally taken when turning; namely, roughing cuts and finishing cuts. In either case, the cutting tool is positioned to cut the required diameter by using the trial-cut procedure. Since it is so frequently used this method will be described again, in greater detail.

Fig. 8-15. Measuring length of workpiece by measuring to layout line with hermaphrodite calipers. These calipers can also be used to scribe layout lines on cylindrical surfaces, as shown.

The cutting tool is first adjusted to take a shallow cut using the cross slide to position the tool. Care must be taken to make sure that this cut will leave the work oversize. The automatic, longitudinal power feed is engaged and a cut, approximately one-half inch in length, is taken. With the lathe spindle stopped, the diameter of the work is measured with micrometer calipers. The carriage is then moved back toward the tailstock in order to position the tool for taking a second cut. Before this cut is taken, the cutting tool is adjusted to cut the workpiece to the desired diameter by moving the cross slide the exact distance required. This distance is one-half the difference between the measured diameter of the trial cut and the desired diameter. The distance that the cross slide is moved can be accurately determined by reading the micrometer cross-feed dial. As explained in Chapter 5, some cross-feed dials are direct reading, while others must be moved one-half of the difference between the trial-cut diameter and the desired diameter. Since direct reading dials have graduations that are equal to a movement of one-half of a thousandth of an inch, this subtraction is not necessary. On direct-reading dials the cross slide is moved the required distance by reading the number of graduations corresponding to the difference in the diameter. It must be understood that in either case, the cross slide moves the same distance, and that the difference is in how the micrometer cross-feed dials are graduated. After the cutting tool has been set to take the second cut, the longitudinal feed is engaged and another one-half-inch-long cut is taken. The resulting diameter is measured. If the diameter of the work is to size, the longitudinal feed is engaged and the cut is taken to the required length. It is sometimes necessary to take several trial cuts before the desired diameter on the work is obtained.

Ch. 8 CYLINDRICAL TURNING 185

The entire workpiece should be "roughed out" before taking any finish cuts by taking one or more roughing cuts on each diameter that is to be turned. The roughing cuts should leave the diameter of the workpiece oversize a predetermined amount, which in general practice, may vary from approximately .020 to .060 inch. Shoulder lengths should be left short, or "undersize," in order to provide metal for the finish-turning operation. This amount is usually, approximately, .015 to .030 inch.

The finishing cut is often taken with the same cutting tool as was used to take the roughing cut. This procedure requires fewer changes of cutting tools. It is sometimes best, however, to use a separate tool for finish turning since less tool wear then occurs. The separate finish-turning tool, having less wear than the roughing tool, will frequently produce a better surface finish. Sometimes special tools, such as shown in Fig. 7-11, are used for the finish-turning operation. The trial-cut procedure is used to obtain the finished diameter on the workpiece. When finish turning, great care must be exercised in the performance of this procedure so that the work diameter is not cut undersize. The final finishing cut should not be taken until the lathe operator is certain that the cutting tool is set to cut the work to the desired diameter.

The cutting speed for finish turning can usually be faster than for rough turning because of the decrease in the depth of cut; however, not all lathes

Courtesy of The R. K. LeBlond Machine Tool Company

Fig. 8-16. Turning and polishing large-diameter workpieces. Note the polishing straps used to polish large diameters on the lathe at the left. These are enclosed in the device that looks like a large nutcracker.

can be adjusted to obtain this speed differential. Cemented-carbide cutting tools should be operated at a cutting speed that is fast enough to produce a good surface finish. When these cutting tools are operated above a certain critical cutting-speed range, an excellent surface finish can easily be obtained. This critical cutting-speed range is dependent on the grade of carbide used and the nature of the work material. In general, it is in the range of 150 to 200 fpm. When slower cutting speeds must be used, the surface finish obtained can be improved by using a good cutting fluid and by using a cutting edge that is in good condition. This usually means that the cutting edge must be replaced before excessive wear has taken place.

Sometimes, filing and polishing are used to obtain a smooth finish on parts that have been turned in a lathe. The file that is generally used for lathe work is a single-cut, bastard file of "mill" section. The amount of metal that is left for filing should be kept to a minimum. It is difficult to file a part round, in a lathe. In general, filing should be avoided except to "break" sharp corners of shoulders. An abrasive cloth may be used to obtain a smooth surface by pressing it against the rotating workpiece. A rather smooth finish can be obtained by using the finer grade of abrasive cloth. Polishing straps are sometimes used when polishing large diameters. A polishing strap is shown in Fig. 8-16. The two ends of the polishing strap are held together with a piece of leather belting.

Figure 8-17 illustrates a digital readout system that can be adapted to engine lathes. This system reduces the time and effort required to set the cutting tool in order to turn the work to the required diameter and length. The cutting tool is set against the end face, and the digital display for the longitudinal feed is set to zero. A short trial cut is taken and the diameter is measured with a micrometer caliper. The digital display for the cross slide is then set to read the diameter of the trial cut. From then on, the position of the point of the cutting tool—with respect to length and diameter—will appear on the digital display. Thus, the lathe operator can set the cutting tool to cut the length and diameter wanted, by observing the dimensions displayed on the digital display without taking any further trial cuts. For accurate work, of course, roughing cuts should be followed by a light finishing cut to minimize the effect of deflections caused by the heavy roughing cuts.

Shoulder Turning

There are three steps in shoulder turning:

1. Locating the shoulder on the work
2. Turning the smaller diameter to size
3. Facing the shoulder to length.

The length of the shoulder can be laid out using a hermaphrodite caliper as shown in Fig. 8-15. A scratch, or mark, made with the point of the cutting tool can also be used to mark the location of the shoulder. Often, the location of the shoulder is measured as the cut progresses, using a

Ch. 8 CYLINDRICAL TURNING 187

Courtesy of The Bendix Corporation, Automation and Measurement Division

Fig. 8-17. Engine lathe equipped with Sheffield Digiturn Lathe-Control System.

steel rule or a hermaphrodite caliper. A groove cut with a necking tool, A, Fig. 8-18, can serve to locate a shoulder.

Several methods of turning shoulders are illustrated in Fig. 8-18. At A, a groove is first cut with a necking tool in order to establish the length of the shoulder. The small diameter is then cut to size by one or more cuts with the turning tool. When this method is used, all of the required shoulders are "blocked out," and the adjacent cylindrical surfaces are cut to size. This method does have a disadvantage, in that two tools need to be used. On slender workpieces a very slow feed is necessary to cut the grooves in order to avoid the occurrence of chatter. A faster, and even simpler method of cutting a shoulder is shown at B. The cutting tool is positioned to have a zero-degree lead angle. The shoulder is formed by the cutting edge when turning the smaller diameter. Shoulders turned in this manner are not necessarily perfectly square. When a shoulder must be square the method illustrated at C, is recommended. Usually the shoulder is first roughed out using either of the two previous procedures. The tool is then reset with the nose leading, as shown at C. It is set to cut the small diameter to size, using the trial-cut procedure. The microm-

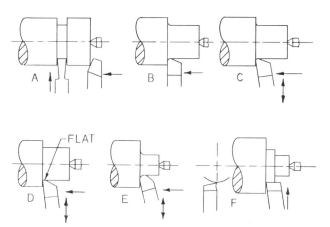

Fig. 8-18. Procedures for turning shoulders.

eter cross-feed-dial reading used to take this cut, is noted. When the tool approaches the shoulder the automatic-power longitudinal feed is disengaged and the nose of the tool is fed by hand, longitudinally, into the shoulder. When it has penetrated the shoulder enough to take a light cut, it is fed out using the cross slide. The length of the shoulder is then measured, and the tool is moved longitudinally, just enough to remove the excess metal from the shoulder. It is then fed into the work with the cross slide. When it approaches the corner of the shoulder it is carefully fed by hand, until the reading of the micrometer cross-feed dial is the same as was previously noted. In this manner the tool will not undercut the corner and a smooth blend can be obtained between the shoulder and the previously turned diameter. It is sometimes an advantage to turn the compound rest 90 degrees, for this operation. In this position, the compound-rest slide will move parallel to the longitudinal feed, and it can be used to set the depth of cut for the final finishing cut on the shoulder. Very accurate shoulder lengths can be obtained using this method of setting the tool. A micrometer carriage stop, Fig. 6-19, if available, can also be used to obtain accurate shoulder lengths. This same procedure can be used to cut a shoulder having a sharp corner, as shown at D, Fig. 8-18, or to cut a radius in the corner, as shown at E. The tool, at D, must be ground with a flat on the end-cutting edge in order to finish turn the small diameter, otherwise it will cut threadlike grooves on this surface. Very large radii are cut using a large forming tool that is similar to the tool at E. The corner is first blocked out using a turning tool, such as shown at B. It is sometimes faster to block out the corner with a facing tool, as shown at F. The corner radius is finished by touching the finished surface at the small diameter without penetrating this diameter with the radius-forming tool. This can be done if care is exercised. The reading of the micrometer cross-feed dial is noted when the radius forming tool is in this position. The radius is then cut, as shown at E, with care being taken

at all times not to move the micrometer cross-feed dial beyond the setting noted above. A heavy-duty cutting oil, or white lead mixed with oil to a thin consistency, should be used as a lubricant when performing this operation in order to obtain a good surface finish in the radius. The cutting speed should also be greatly reduced when using the large radius form tool.

Mandrels

A mandrel is a work-holding device that is pressed into a finished machined hole of a workpiece in order to rotate the workpiece accurately about the axis of the hole, while the mandrel is held between the lathe centers. Standard mandrels are made from hardened and ground steel. They have a taper of .006 inch per foot. The size of standard mandrels is always marked on the large end. A thin film of oil should be placed on the mandrel before it is inserted into the hole. The small end of the mandrel should be inserted into the hole and the mandrel pressed in until it is tight. If hammer blows are used for this purpose, the end of the mandrel should be protected with a piece of soft metal or hard wood. The best practice is to insert the mandrel into the hole with an arbor press.

Two examples of the application of mandrels are shown in Fig. 8-19. In the upper view, the outside diameter of the bushing is turned to size and is concentric with the bore that has previously been finished to size. In the lower view, the small cast-iron wheel, having a finished hole

Fig. 8-19. (Upper Diagram) Bushing mounted on a mandrel for turning. (Lower Diagram) Turning a pulley held on a mandrel.

Courtesy of the Midwest Industrial Management Association

Fig. 8-20. A. Nut mandrel, B. Gang mandrel, C. Cone mandrel, D. Expanding mandrel.

through the hub, is having the outside diameter turned concentric and the sides faced perpendicular to the axis of the hole. If a comparatively large pulley is mounted on the mandrel, it must be driven directly by pins attached to a faceplate and engaging the pulley arms in order to prevent the pulley from turning on the mandrel when taking a cut.

Some special mandrels are shown in Fig. 8-20. A nut mandrel is shown at A. Often, this type of mandrel is made by chucking a piece of scrap stock and turning it to the size required to fit the threads on the work. Thus, the threads are concentric with the rotation of the lathe spindle and any thread mounted on this mandrel will likewise rotate concentrically with the lathe. Nuts and other threaded workpieces can have surfaces turned concentrically, and faces turned perpendicularly, with respect to

the threads that are held by the nut mandrel. Gang mandrels, shown at B, can be used to hold two or more parts between centers in a lathe. Cone mandrels can be used to hold large diameters. A cone mandrel can be seen at C. Expanding mandrels, D, consist of a slotted sleeve that has a tapered hole which fits onto a tapered shaft. When the shaft is forced onto the sleeve, the taper forces the sleeve to expand and to tighten inside the hole of the workpiece.

Steady Rests and Follower Rests

One of the most useful lathe accessories is the steady rest shown in Fig. 8-21. The purpose of the steady rest is to support long, slender parts, in a lathe. The additional support given by the steady rest reduces the deflection of the workpiece caused by cutting forces; thereby, the accuracy of the work is improved and the tendency for chatter to occur is reduced. On very long and slender workpieces a steady rest is sometimes required just to keep the workpiece from sagging under its own weight. The steady rest may be used to support workpieces that are held between centers (Figs. 8-16 and 8-21), with one end held in a chuck and the other on the tailstock center (Fig. 8-2, tailstock center not shown), and with one end held in a chuck and the other supported by the steady rest only (Fig. 8-6) in order to leave the end face open for machining.

The steady rest is composed of a frame containing three movable jaws which form a three-point bearing. The frame is hinged on one side to allow the upper half to swing back for inserting or removing the work. The base of the frame fits on the ways of the lathe bed and is secured to

Courtesy of The R. K. LeBlond Machine Tool Company

Fig. 8-21. Steady rest used to support a long, slender part that is held between centers.

Courtesy of The R. K. LeBlond Machine Tool Company

Fig. 8-22. Follower rest used to support long, slender workpieces that must be turned along entire length.

the bed with a clamp. Before applying the steady rest, it may be necessary to turn a true surface, or spot, on the workpiece in order to provide a bearing surface for the jaws of the steady rest. The spot is turned by a tool having a small nose radius and a 0-degree lead angle. The spot is sometimes polished with an abrasive cloth. The jaws should be carefully adjusted against the workpiece. For very accurate work, a dial test indicator is positioned on the workpiece 180 degrees opposite one of the lower jaws. This jaw is brought against the work until the indicator shows a deflection of a few-thousandths-of-an-inch. The jaw is then carefully backed off until the indicator returns to its original reading. The second lower jaw is adjusted by the same procedure. The top jaw is then brought down on the work and adjusted so that the work will turn freely, but without play.

The surface against which the jaws are brought to bear should be kept well oiled. If the jaws are required to bear against a finished surface, this surface should be protected by placing a strip of leather belting, or a strip of abrasive cloth (with the abrasive side out) between the jaws and the work. The steady rest must be located where it will not interfere with the turning tool. In some cases this may require the steady rest to be moved to another position after turning one section.

Sometimes, a long cylindrical surface must be turned on a workpiece so that there is no room to place the steady rest without interfering with the cut. In such instances a follower rest, Fig. 8-22, must be used to support long, slender workpieces against deflections. The follower rest is similar to the steady rest except that it has only two jaws and it is

attached to the carriage. The jaws are adjusted to the finished surface after a short cut has been taken. The follower rest then moves along behind the cutting tool for the remaining length of the cut.

Knurling

A knurled surface is a raised surface, usually with a diamond-shape pattern, which is formed to produce an ornamental effect or to provide a nonslip or gripping surface. This pattern is formed on the surface by a knurling tool such as is shown in Fig. 8-23. This tool has three pairs of rolls which can be made to produce a fine, medium, or a coarse-pitch knurled surface. The rolls, or knurls, are made from hardened steel and each knurl has a series of teeth or ridges that incline to the right on one knurl and to the left on the other. When the two knurls are pressed into the work as it slowly rotates, one knurl forms a series of left-hand ridges and the other a series of right-hand ridges which cross, to form the diamond-shape pattern. The knurled surface is formed by the high local pressures produced by the knurls which cause the metal on the surface of the work to flow, or extrude, in an action which is similar to center punching. Knurled surfaces with a single groove are sometimes made, although the diamond pattern is the most common on machine parts. Several different kinds of knurling tools are also used. Some knurling tools are designed to reduce the load, which can be very high, on the cross-feed screw and nut.

The knurling operation is performed by clamping a knurling tool in the toolholder and aligning it so that the knurls are parallel to the surface of the workpiece. The lathe should be set to operate at a very slow speed

Courtesy of The R. K. LeBlond Machine Tool Company

Fig. 8-23. Knurling to provide a slip-free surface.

and a plentiful supply of lubricating oil should be placed on the surface to be knurled. The knurling operation should be started, if possible, with about one-third or one-half of the knurl in contact with the surface to be knurled. The knurl is then fed inward, using the cross feed, until a well-formed knurled surface is visible on the surface of the work. If a fine knurl is being made it may be desirable, at this point, to feed the knurl into the full depth. On harder materials and coarse knurl patterns, the initial feed should not be to the full depth unless the width of the knurled surface on the work is less than the knurl rolls, in order to prevent an excessive load from occurring on the cross-feed screw and nut. If the surface to be knurled is wider than the knurls, the power longitudinal feed of the lathe should be engaged and the knurling tool is traversed back and forth, feeding it inward before each passage, until the full depth is reached. The knurls should never be allowed to lose contact with the work until the job is finished. The correct depth is obtained by observing the pattern of the knurl on the part.

Horsepower Requirements for Turning

The horsepower required to turn metal is dependent upon the cutting speed, the depth of cut, the feed rate, and by the unit horsepower of the material being cut. The unit horsepower is an experimentally determined value. It is the horsepower required to cut a material at a rate of one

Table 8-1. Unit Horsepower for Ferrous Metals and Alloys

Material	Brinell Hardness Number (Bhn)					
	150 to 175	176 to 200	201 to 250	251 to 300	301 to 350	351 to 400
	Unit Horsepower, uhp					
AISI 1010–1025 Steel	.58	.67
1030–1055	.58	.67	.80	.96
1060–109575	.88	1.0	...
1112–1120	.50
1314–1340	.42	.46	.50
1330–135067	.75	.92	1.1	...
2015–2115	.67
2315–2335	.54	.58	.62	.75	.92	...
2340–235050	.58	.70	.83	1.0
2512–2515	.50	.58	.67	.80	.92	...
3115–3130	.50	.58	.70	.83	1.0	...
3160–345050	.62	.75	.87	1.0
4130–434546	.58	.70	.83	1.0
4615–4820	.46	.50	.58	.70	.83	.87
5120–5150	.46	.50	.62	.75	.87	1.0
5210058	.67	.83	1.0	...
6115–6140	.46	.54	.67	.83	1.0	...
6145–619570	.83	1.0	1.2	1.3
Gray Cast-Iron	.30	.33	.42	.50
Alloy Cast-Iron	.30	.42	.54
Ferritic Malleable Iron	.42
Pearlitic Malleable Iron50	.54	.75
Cast Steel	.62	.67	.80

Table 8-2. Unit Horsepower for Nonferrous Metals and Alloys

Material	Bhn	uhp	Material	Bhn	uhp
High-Temperature Alloys			Zinc Alloy (die cast)		.25
			Monel (rolled)		1.0
A286	165	.82	Brass:		
A286	285	.93	Hard		.83
Chromoloy	200	.78	Medium		.50
Chromoloy	310	1.18	Soft		.33
Hastalloy-B	230	1.10	Free Machining		.25
Inco 700	330	1.12	Bronze:		
Inco 702	230	1.10	Hard		.83
M-252	230	1.10	Medium		.50
M-252	310	1.20	Soft		.33
Ti-150A	340	.65	Copper (pure)		.90
U-500	375	1.10	Aluminum		
AISI 4340 Steel	200	.78	Cast		.25
AISI 4340 Steel	340	.93	Hard (rolled)		.33

cubic inch per minute. Values of unit horsepower for different metals are given in Tables 8-1 and 8-2. Table 8-3 gives the correction factor for feed.

The formula for calculating the horsepower at the cutting tool is given below:

$$\text{hp}_c = \text{uhp}\, C\, 12\, Vfd \qquad (8\text{-}1)$$

where: hp_c = Horsepower at the cutting tool
uhp = Unit horsepower or the horsepower required to cut metal at a rate of one cubic inch per minute
C = Feed correction factor
V = Cutting speed, feet per minute
f = Feed, inches per minute
d = Depth of cut, inches

Table 8-3. Feed Correction Factor, C, for Unit Horsepower (See Formula 8-1)

Feed, ipr	Feed, Correction Factor, C	Feed, ipr	Feed, Correction Factor, C
.001	1.65	.021	.89
.003	1.32	.025	.86
.005	1.16	.030	.83
.007	1.12	.035	.81
.010	1.04	.040	.79
.012	1.00	.050	.75
.015	.96	.060	.72
.018	.92	.090	.67

The horsepower required to cut the metal can be calculated by Formula 8-1; however, this is not all of the power that must be supplied by the motor. Additional power is required to overcome the mechanical resistances in the lathe, such as bearing friction. The power load imposed by

the lathe must be added to the power required by the cutting tool in order to determine the horsepower that the electric driving motor must supply. The total horsepower that must be supplied by the driving motor can be estimated by using a mechanical efficiency factor. If the lathe had no mechanical power losses its mechanical efficiency would be considered to be 1.0, and all of the horsepower supplied by the driving motor is available at the cutting tool to take the cut. If only 90 per cent of the horsepower supplied by the driving motor is available for cutting, the remaining 10 per cent is required to overcome the mechanical resistances, and the mechanical efficiency of the lathe would be considered to be .9.

The total horsepower required to drive the lathe is then determined by Formula 8-2:

$$\text{hp}_m = \frac{\text{hp}_c}{e_m} \qquad (8\text{-}2)$$

where: hp_c = Horsepower at the cutting tool
hp_m = Horsepower at the driving motor
e_m = Mechanical efficiency factor
e_m = .90 for direct spindle drive with a belt
e_m = .75 for back gear drive
e_m = .70 to .80 for geared head drives.

(*Note:* There are some drives that have an efficiency as low as .40. The above values represent average values.)

The horsepowers calculated by Formulas 8-1 and 8-2 are valid when using sharp cutting tools. When the cutting tools are dull, more horsepower will be required; this may be as much as two times the calculated values if the tools are very dull.

Example 8-1:

Calculate the horsepower used by the cutting tool and by the lathe driving motor in order to turn a piece of AISI 1020 steel with a high-speed cutting tool. The cutting speed to be used is 130 fpm, the feed is .010 inch per revolution, the depth of cut is .125 inch, and the hardness of the steel is 180 Bhn. The lathe spindle is driven by a direct-belt drive.

$$\text{uph} = .67 \text{ (from Table 8-1)}$$

$$C = 1.04 \text{ (from Table 8-3)}$$

$$e_m = .9$$

$$\begin{aligned}\text{hp}_c &= \text{uph } C \; 12 \; V \; f \; d \\ &= .67 \times 1.04 \times 12 \times 130 \times .010 \times .125\end{aligned}$$

$$\text{hp}_c = 1.36$$

$$\text{hp}_m = \frac{\text{hp}_c}{e_m}$$

$$= \frac{1.39}{.9}$$

$$\text{hp}_m = 1.5$$

Example 8-2:

Calculate the horsepower required to turn a shaft that is made from AISI 4340 steel, with a hardness of 240 Bhn, on a geared-head lathe. The feed rate to be used is .020 inch per revolution and the depth of cut will be .125 inch. The cutting tool material will be cemented carbide.

$V_o = 325$ fpm (from Table 7-1)

$F_f = .74$ (from Table 7-2)

$F_d = 1.00$ (from Table 7-2)

uhp $= .58$ (from Table 8-1)

$C = .89$ (from Table 8-3)

$e_m = .80$

$V = V_o F_f F_d$
$= 325 \times 0.74 \times 1.0$

$V = 240$ fpm

$\text{hp}_c = \text{uhp } C \; 12 \; V f d$
$= .58 \times .89 \times 240 \times .020 \times .125$

$\text{hp}_c = 3.72$

$$\text{hp}_m = \frac{hp_c}{e_m}$$

$$= \frac{3.72}{.80}$$

$\text{hp}_m = 4.65$

Example 8-3:

Calculate the maximum feed that can be used when cutting AISI 1040 steel which has a unit horsepower (uhp) of .67. The depth of cut is to be .250 inch. The cutting tool material is high-speed steel and the cutting speed to be used is 90 fpm. The lathe to be used has a 3-horsepower motor and its mechanical efficiency is .75.

This problem cannot be solved directly because the C, or feed correction factor, must be estimated in the initial trial. By successive trials a C factor can be estimated which will closely correspond to the answer.

$$\text{hp}_c = \text{uhp } e_m \quad \text{(from Formula 8-2)}$$
$$= 3 \times .75$$
$$= 2.25$$

(Thus, there are 2.25 horsepower available at the cutting tool in order to take the cut.)

$C = .86$ (Estimated from Table 8-3 for an assumed .025 ipr feed rate.)

$$\text{hp}_c = \text{uhp } C\, 12\, V\, f\, d$$

Solving this formula for f:

$$f = \frac{\text{hp}_c}{\text{uhp } C\, 12\, V\, d} = \frac{2.25}{.67 \times 86 \times 12 \times 90 \times .250}$$
$$= .014 \text{ ipr}$$

The estimated C factor is too low since it is for a feed rate of .025 ipr. (Note that the value of the C factor decreases with an increasing feed.) Thus, a new, higher C factor must be estimated that will result in a lower permissible feed.

$C = 1.00$ (Second estimated C factor for an assumed .012 ipr feed rate.)

$$f = \frac{\text{First } C \text{ factor}}{\text{Second } C \text{ factor}} \times \text{Feed calculated using first } C \text{ factor}$$

$$= \frac{.86}{1.00} \times .014$$

$$f = .012 \text{ ipr}$$

Thus, the maximum feed that can be used for the given cutting conditions and using all of the available rated horsepower of the motor would be .012 inch per revolution. Actually, a somewhat higher feed rate could be used but if the higher feed is used for a prolonged period of time the motor would overheat.

QUESTIONS

8-1. Name the five basic turning operations.
8-2. Describe four different methods of holding a workpiece in a lathe.
8-3. Describe three methods of drilling the center holes in the ends of cylindrical workpieces.
8-4. Why must the headstock center of the lathe run true?
8-5. How is the headstock center of the lathe trued?
8-6. What is the difference between a live, and a dead, tailstock center?

Ch. 8 CYLINDRICAL TURNING 199

8-7. When is a live tailstock center used?

8-8. Why must the tailstock center be aligned to coincide with the lathe axis?

8-9. How is the tailstock center made to coincide with the lathe axis?

8-10. How is the workpiece set up in the lathe for turning between centers?

8-11. In what position should the cutting tool be held with respect to the lathe centers?

8-12. How is the feed rate of the lathe determined?

8-13. Describe the procedure for facing the ends of a cylindrical shaft that is held between centers in a lathe.

8-14. What are the advantages of turning a part between centers?

8-15. Describe the procedure for setting a single-point lathe tool to reduce the diameter of a shaft from 1.000 to .875 inch in one cut.

8-16. What are the three basic steps in turning shoulders?

8-17. Describe three different methods of turning a shoulder.

8-18. What is a mandrel? When and why is it used?

8-19. What is a nut mandrel?

8-20. How is a steady rest adjusted accurately to a workpiece that is held on lathe centers?

8-21. When should a follower rest be used?

8-22. What is meant by knurling?

8-23. What is meant by unit horsepower?

8-24. Calculate the horsepower required by the cutting tool and by the lathe motor in order to turn a casting made from a Class 30 gray cast-iron having a hardness of 180 Bhn using a feed of .015 ipr and a depth of cut of .125 inch. The cutting tool is made from high-speed steel and the cutting speed to be used is 85 fpm. The efficiency of the lathe is .8.

8-25. Calculate the horsepower required to turn a shaft that is made from AISI 5150 steel having a hardness of 250 Bhn. The depth of cut to be used is .250 inch and the feed is .030 inch per revolution. The cutting-tool material is cemented carbide. The efficiency of the lathe is .7.

8-26. Calculate the maximum feed that will utilize all of the available horsepower at the lathe when cutting AISI 1020 steel having a hardness of 160 Bhn. The lathe has a one horsepower driving motor and an efficiency of .8. The depth of cut is to be .125 inch. The cutting-tool material is high-speed steel and the cutting speed is to be 110 fpm.

CHAPTER **9**

Chucking Work

There are many different designs and types of chucks used in a machine shop. For example, there are drill chucks for holding straight-shank twist drills, magnetic chucks, vacuum chucks, etc. The four-jaw independent chuck, the three-jaw universal chuck, and collet chucks are commonly used in the engine lathe to clamp workpieces and hold them in alignment for the performance of chucking operations which includes turning, facing, thread cutting, drilling, boring, reaming, and cut off. Short workpieces are held only by the chuck, while longer ones are given additional support by the tailstock center or by a steady rest.

Mounting the Workpiece in a Three-Jaw Universal Chuck

A detailed view of a three-jaw chuck is shown in Fig. 9-1. The workpiece is placed in the chuck jaws and the jaws are closed by turning bevel pinion C, with the chuck key. The bevel pinion engages the gear teeth on the back of the scroll B, which turns inside of the chuck body D. The

Courtesy of the Buck Tool Company

Fig. 9-1. Details of a 3-jaw, Adjust-Tru, universal scroll chuck.

chuck jaws A, engage the threads on the scroll. When it rotates, the jaws move in the slides that are machined onto the face of the chuck body. Each slide and jaw is numbered, for each jaw must be assembled into its corresponding slide by always starting with jaw number 1, then number 2, and finally, number 3. The back plate, or mounting plate, F, contains the mounting used to hold the chuck on the lathe spindle nose. In this illustration a threaded mounting is shown.

Because of the clearances required to allow this mechanism to operate without binding and because of wear, three-jaw chucks do not always hold the workpiece accurately in terms of a few thousandths of an inch. In many cases this is good enough; however, when greater accuracy is required the jaws must be ground true while in the chuck body; a difficult procedure to perform. This can be avoided with the chuck shown in Fig. 9-1. A fine adjustment of the chuck is provided that is independent of the chuck jaws. Four opposite socket-head screws, G, work on pressure plugs, H, that contact surface, F, of the back plate in order to shift the complete chuck and work into alignment. The work is placed in the chuck, Fig. 9-2, and tightened with the chuck key by turning pinion C. A dial test indicator is used to check the concentricity of the work. Any eccentricity can be rapidly removed by adjusting opposite pairs of socket-head screws, G. No further adjustment of the socket screws is required when chucking duplicate parts.

Mounting the Work in a Four-Jaw Independent Chuck

The four-jaw independent chuck is a very versatile chuck with a wide range of applications on engine lathe work. Each jaw works independently by means of a separate screw turned by the chuck key. Parts can be held firmly and, if required, very accurately, in this chuck. The accuracy with which a part can be held is dependent entirely on the condition of the surface on the work that is used to test the concentricity. A rough-cast surface that is not round, obviously cannot be chucked to "thousandths," because no reference surface exists to prove the truth of the position of the part. Cold-drawn and machined surfaces, on the other hand, can be indicated into position within "thousandths," using a dial test indicator as shown in Fig. 9-4. When the procedure is understood, four-jaw chucks can usually be adjusted accurately within minutes.

The chuck jaws should be opened just far enough to admit the workpiece by observing the position of each jaw in relation to concentric grooves that are machined into the face of the body. The jaws are then closed to hold the work lightly. The shank end of a tool holder, or a piece of metal that is resting in the tool post, is placed against the work—opposite one of the chuck jaws—as shown at A, Fig. 9-3. The tool holder must not be clamped tightly in the tool post as it must be able to be pushed aside with little effort. The chuck is then turned manually. The high spot on the work pushes the shank of the tool away resulting in an opening, D (shown in B, Fig. 9-3). This occurs when the low spot is opposite the shank. All effort should first be made to adjust the jaw with

202 CHUCKING WORK Ch. 9

Courtesy of the Buck Tool Company

Fig. 9-2. Making a fine adjustment to remove eccentricity on workpiece on a 3-jaw, Adjust-Tru, universal scroll chuck.

the largest opening between the shank and the work, and its opposing jaw, until the openings opposite these two jaws are equal. The other two jaws

Fig. 9-3. Procedure for adjusting workpiece in a 4-jaw independent chuck.

then are adjusted likewise. Finally, the fine adjustment is made again, first by adjusting two opposing jaws, then the other pair of opposing jaws. When the reference surface used to check the concentricity is a finished surface, it should finally result in work that can be rotated without any light showing between this surface and the shank. For very accurate work a dial test indicator can be used in place of the tool shank in making the fine adjustment. (See Fig. 9-4.) The dial test indicator should not be

Fig. 9-4. Using a dial test indicator to test the concentricity of a workpiece held in a 4-jaw independent chuck.

used to check a rough surface. The contact point of the indicator should be used only against a finished surface. Sometimes a hole in a workpiece must be made to run true in a chuck. This is done by indicating inside the hole, as shown in Fig. 9-5. In this case the indicator is held by a magnetic base.

Some parts held in a chuck will tend to show endwise eccentricity or wobble in addition to radial eccentricity. In some cases this can be corrected by placing the work against the face of the chuck jaws or chuck body. When this cannot be done, the shank of the tool holder, or a piece of steel, is used against the face in a similar manner as for centering the work radially. In this case the high spots are found which are then tapped back with light blows from a soft hammer. The fine adjustment can be checked by using a dial test indicator. The contact point of the indicator should not be in contact with the workpiece when it is being tapped into adjustment with the hammer, in order to avoid damage to

Courtesy of The L. S. Starrett Company

Fig. 9-5. Using a dial test indicator to test the concentricity of a hole located in a workpiece held in a 4-jaw chuck.

the indicator mechanism. Very large four-jaw chucks have radial T-slots spaced between the chuck jaws. Studs which extend an equal distance from the face of the jaws can be fastened in the T-slots, and the parts being chucked can be positioned against the end of these studs, thereby obtaining endwise alignment.

The workpiece is sometimes deliberately mounted eccentrically in the four-jaw chuck in order to turn a surface which is eccentric with respect to another surface. (See Fig. 9-6.) The flexibility of the four-jaw chuck is clearly demonstrated in this operation. It is, perhaps, the most versatile and useful work-holding device for engine-lathe work.

Mounting Work in Collet Chucks

Spring collet chucks are highly accurate chucks. They should be used to clamp on finished surfaces that are accurate to within a few thousandths of an inch of the size of the collet used. The spring collets exert an even and uniform pressure around the workpiece, which minimizes the distortion of thin-walled parts. Figure 9-7 shows a group of spring collets used to hold round, square, and hexagon-shape workpieces. A draw-in, bar-type collet chuck is shown in Fig. 9-8. A special collet adapter is

Ch. 9 CHUCKING WORK 205

Courtesy of the Clausing Division of the Atlas Press Company

Fig. 9-6. Work held eccentrically in a 4-jaw independent chuck.

Courtesy of Hardinge Brothers Inc.

Fig. 9-7. Spring collets for holding different work shapes.

Courtesy of South Bend Lathe

Fig. 9-8. Draw-in bar type collet chuck.

inserted in the spindle nose and the collet is inserted in the adapter. The draw-in bar is inserted in the rear of the spindle. When the draw-in bar is turned, it draws the collet into the spindle by the action of the threads. The taper on the adapter (corresponding to the taper on the collet) causes the collet to close as it is pulled into the spindle by the draw-in bar. All of the surfaces of the spindle nose, the adapter, and the collet must be clean since even the smallest particle of dirt or foreign substance will cause the workpiece positioned inside the collet to be held inaccurately. A spindle-nose type collet chuck is shown in Fig. 9-9. This chuck can be obtained to fit onto either a threaded cam lock or a long-taper spindle nose. The collet is inserted in the front of the chuck and pulled back against the taper to permit closing the jaws by turning the handwheel. As it is not necessary to reach behind the spindle to close the chuck, a saving of time and effort results.

Collet-Chuck Work

Collet chucks are used to hold finished surfaces only. This, however, can include raw cold-drawn or centerless ground bar stock. The bar stock can be placed inside of the headstock spindle with enough material protruding outside of the collet to make the required part. The finished part is then cut off with a cut-off tool. Collet chucks are frequently used for second chucking operations. The part is first finish-turned on one end. It is then turned around and the finished end is held in the collet chuck while the other end is turned to size. A turning opeartion with the part held in a spindle-nose type collet chuck is shown in Fig. 9-10. Note that the compound rest is set at 90 degrees so that the compound-rest slide moves parallel to the axis of the lathe. This permits the micrometer dial

Ch. 9 CHUCKING WORK 207

of the compound-rest feed screw to be used to obtain an accurate distance between the shoulder and the end face of the workpiece.

In Fig. 9-11 a very long, small-diameter workpiece is being turned while held in a collet chuck. Turning this between centers would be difficult be-

Courtesy of Hardinge Brothers Inc.

Fig. 9-9. Spindle-nose type collet chuck.

Courtesy of Hardinge Brothers Inc.

Fig. 9-10. Turning and facing a part held in a spindle-nose type collet.

cause the action of the cutting forces would tend to bend the part and the lathe dog would interfere with the cut. The first operation is to hold

Courtesy of the Clausing Division of the Atlas Press Company

Fig. 9-11. Turning a slender workpiece that is held in a collet chuck and given additional support by a follower rest and a steady rest.

Ch. 9 CHUCKING WORK 209

the workpiece inside of the spindle with just enough protruding from the collet to take a cut of about one or two inches in length. It is then drawn out of the collet and one end is held by a steady rest as shown. This end could be held in a tailstock center; however, on very small diameter work the center hole in the workpiece would be very small affording only a small bearing surface for the tailstock center. A follower rest is placed on the carriage which, in this case, leads the cutting tool. The follower rest prevents the work from deflecting as the tool cuts the length of the part to size. This job could also be done by turning the length between the collet chuck and the steady rest to size, as shown, after which the two ends are cut off.

Chucking for Work-Turning and Facing

Cylindrical surfaces can be turned with the work held in the chuck as shown in Fig. 9-12. The longitudinal feed causes the cutting tool to travel parallel to the axis of the lathe and as the work revolves, the tool will cut a true cylindrical surface. The trial-cut procedure is used to obtain the desired diameter on the workpiece. Long parts held in a chuck may

Courtesy of the Clausing Division of the Atlas Press Company

Fig. 9-12. Turning a cylindrical surface with a part that is held by a 3-jaw universal chuck.

deflect as a result of the cutting forces which cause the part to be tapered instead of cylindrical. This can be prevented by using a cutting tool having a small-nose radius and setting this tool to have a zero-degree lead angle.

Facing is the operation of turning a plane surface by feeding the cutting tool with the cross slide. A facing operation is illustrated in Fig. 9-13. The depth of cut for this operation can be obtained by moving the carriage longitudinally.

Greater accuracy in setting the depth of cut can be obtained by setting the compound-rest slide so that it will move parallel to the lathe axis as shown in Fig. 9-13. The depth of cut is set by moving the compound-rest slide using the compound-rest micrometer to make accurate settings. If

Fig. 9-13. Taking a facing cut on a gear blank.

desired, the compound rest can be set at 60 degrees with respect to the lathe axis in order to make the depth-of-cut settings for facing operations. This results in a relationship of the angles such that a movement of .001 inch of the compound-rest slide will move the cutting tool .0005 inch toward (or away from) the face of the workpiece. Thus, if it is desired to set the tool to cut .040 inch deeper when facing, the compound rest is moved .080 inch.

Drilling

The first step in drilling a hole on a lathe is to start the hole with a combination drill and countersink, or center drill, as seen in Fig. 9-14. The center drill is held in a drill chuck which is mounted in the tailstock spindle. The work is rotated by the headstock spindle and a high spindle-speed should be used to drill this starting hole. The hole is drilled by clamping the tailstock to the bed of the lathe and feeding the drill into the work manually by turning the tailstock handwheel. The center drill, being short and inflexible, will not deflect; but will start the hole true, without any eccentricity. The twist drill will be guided by the hole made by the center drill and, therefore, will also start true and without eccentricity.

After the starting hole has been made with the center drill, the hole is then drilled to the required size and depth with a twist drill. The spindle speed used to drill the hole should be appropriate both to the size

Fig. 9-14. Starting a drilled hole with a center drill.

of the drill and the material being cut. As a guide, the Table of Cutting Speeds (in the Appendix) can be used. The spindle speed is calculated from the cutting speed using Formula 7-2a. The drill is fed into the work manually, by turning the tailstock handwheel.

Small twist drills (up to approximately one-half inch in diameter) can be held in a drill chuck which is mounted in the tailstock spindle, as shown in Fig. 9-15. The taper locates the drill on the axis of the lathe and it will hold the drill sufficiently firm to resist the drilling torque. The torque, encountered when larger drills are used, may be sufficiently large to cause the taper shank to slip in the taper hole of the tailstock spindle. This will result in scoring the inside of the taper hole of the spindle. This damage is frequently so severe that the tailstock spindle is ruined. It is, not ad-

Courtesy of The R. K. LeBlond Machine Tool Company

Fig. 9-15. Drilling a small-diameter hole with a drill held in a drill chuck which is mounted in the tailstock spindle.

visable, therefore, to place large, taper-shank drills directly into the tailstock spindle unless some means are provided to prevent the drill from slipping or some methods used to reduce the drill torque.

Courtesy of The R. K. LeBlond Machine Tool Company

Fig. 9-16. Drilling a large-diameter hole with a taper shank twist drill. The lathe dog prevents the drill from slipping.

The tailstock spindle on some lathes is provided with a keyway, or slot, into which the tang of the twist drill fits. On lathes which are so equipped large-size taper-shank drills can be safely driven by the tang. Most engine lathes do not have a driving slot in the tailstock spindle, thus other means must be used to drive the larger-size twist drills. One method of doing this is shown in Fig. 9-16. A lathe dog is clamped onto the twist drill which is inserted in the tailstock spindle so that the tail of the dog rests on the compound rest. The drill should first be inserted in the spindle after which the dog is clamped on the drill so that the tail is against the compound rest. The tail of the dog prevents the drill from turning. Care must be exercised to make certain that the tail of the dog is against the compound rest at all times as the hole is being drilled.

Another device used in driving large-diameter twist drills is a drill holder which holds the drill outside the tailstock spindle (see Fig. 9-17). The taper shank of the twist drill fits into a matching taper inside the drill holder which is supported by a tailstock center and by the workpiece. The arm of the drill holder is placed on the shank of the tool holder and against the back of the tool post. As the drill is fed through the workpiece by the tailstock spindle the arm of the drill holder pushes the carriage toward the headstock. The carriage thus provides a drag, or a constraint to the movement of the drill holder, thereby keeping it seated on the tailstock spindle at all times. This constraint is especially important in overcoming the tendency of the drill to pull into the work as it breaks through the far side of the workpiece.

Large holes can be safely drilled with taper-shank drills mounted directly in the tailstock spindle by using the "step drilling" procedure. A small hole, say, one-half inch in diameter, is drilled first. This is followed by enlarging the hole with successively larger-diameter drills until the required size is reached. In this way the torque load encountered by any one drill is not excessive, and the drill will not slip in the tailstock

Fig. 9-17. Drilling a large-diameter hole in a lathe. The twist drill is held in a drill holder.

spindle. Another advantage of this method is that the thrust load is reduced which, in turn, reduces the load on the tailstock spindle nut and screw. This may be significant when drilling on a small lathe. The single disadvantage of employing this method is that more time will be required to drill the hole.

Reaming

Reamers are used to obtain an accurate diameter as well as a good surface finish in holes that are being machined in a lathe. Usually the hole is bored prior to reaming in order to make certain that the resulting hole will be concentric with the rotation of the work. The reamer will follow the existing hole through which it cuts. The reamer is a finishing tool and is designed to remove only a small amount of metal. Approximately 0.008 to 0.035 inch should be left on the diameter of the hole for reaming. In other words, the hole should be 0.008 to 0.035 inch smaller in diameter than the size of the reamed hole. In order to prevent a tapered hole resulting from reaming, it is best to hold the reamer in a floating reamer

Courtesy of the Midwest Industrial Management Association

Fig. 9-18. Reaming a hole in a lathe. The reamer is held in a floating reamer holder.

holder as shown in Fig. 9-18. This tool allows a small amount of unobstructed movement which enables the reamer to accurately align itself to ream a straight hole. The cutting speed for reaming should be about one-half of the speed recommended for drilling, and the reamer should be fed through the hole slowly. On engine lathes the reamer is fed manually through the hole by using the tailstock spindle. Cast iron is usually reamed dry; however, a good grade of cutting oil should be used for reaming most other metals including all of the steels.

Boring

The objective of boring a hole in a lathe is:

1. To enlarge the hole
2. To machine the hole to the desired diameter
3. To accurately locate the position of the hole
4. To obtain a smooth surface finish in the hole.

The motion of the boring tool is parallel to the axis of the lathe when the carriage is moved in the longitudinal direction and the workpiece revolves about the axis of the lathe. When these two motions are combined to bore a hole, it will be concentric with the axis of rotation of the lathe. The position of the hole can be accurately located by holding the workpiece in the lathe so that the axis about which the hole is to be machined coincides with the axis of rotation of the lathe. When the boring operation is done in the same setup of the work that is used to turn and face it, practically perfect concentricity and perpendicularity can be achieved.

The setup for boring a hole in an engine lathe is shown in Fig. 9-19. The boring tool is held in a boring bar which is fed through the hole by the carriage. The boring tool should not have a large lead angle. Fre-

Courtesy of the Clausing Division of the Atlas Press Company

Fig. 9-19. Boring a hole in an engine lathe.

quently, it is best to make the lead angle equal to zero degrees in order to minimize the deflection of the boring bar. Also, the nose radius of the boring tool should not be too large. The cutting speed used for boring can be equal to the speed used for turning; however, when calculating the spindle speed of the lathe, the finished, or largest, diameter should be used. The feed rate for boring must usually be less than that used for turning because of the lack of rigidity of the boring bar.

The boring operation is generally performed in two steps; namely, rough boring and finish boring. The objective of the rough-boring operation is to remove the excess metal rapidly and efficiently, and the objective of the finish-boring operation is to obtain the desired size, surface finish, and location of the hole. The size of the hole is obtained by using the trial-cut procedure. The diameter of the hole can be measured with inside calipers and outside micrometer calipers as described in Chapter 2, Basic Measuring Instruments, or inside micrometer calipers can be used to measure the diameter directly.

Cored holes and drilled holes are sometimes eccentric with respect to the rotation of the lathe. When the boring tool enters the work, the boring bar will take a deeper cut on one side of the hole than on the other, and will deflect more when taking this deeper cut, with the result that the bored hole will not be concentric with the rotation of the work. This effect is corrected by taking several cuts through the hole using a shallow depth of cut. Each succeeding shallow cut causes the resulting hole to be more concentric than it was with the previous cut. Before the final, finish cut is taken, the hole should be concentric with the rotation of the work in order to make certain that the finished hole will be accurately located.

Shoulders, grooves, contours, tapers, and threads are also bored inside of holes. The procedure for boring tapers and threads will be described in subsequent chapters. Internal grooves are cut using a tool that is similar to an external grooving tool (see Fig. 8-1). The procedure for boring internal shoulders is very similar to the procedure for turning shoulders described in Chapter 8, Cylindrical Turning. Large shoulders are faced with the boring tool positioned with the nose leading, and using the cross slide to feed the tool. Internal contours can be machined using a tracing attachment on a lathe as shown in Fig. 9-20. The tracing attachment is mounted on the cross slide and the stylus, 1, in Fig. 9-20 follows the outline of 2, the master profile plate. This causes the cutting tool to move in a path corresponding to the profile of the master profile plate in a manner explained previously in Chapter 6, Engine-Lathe Construction. Thus, the profile on the master profile plate is reproduced inside the bore. The master profile plate is accurately mounted on a special slide which can be precisely adjusted in two directions, in order to align the cutting tool in the correct relationship to the work. This lathe has a cam-lock type of spindle nose which permits it to take a cut when rotating in either direction. Normal turning cuts are taken with the spindle rotating counterclockwise. In Fig. 9-20, the boring cut is taken with the spindle revolving in a clockwise direction, or "backwards." This permits the boring cut to

Courtesy of The Monarch Machine Tool Company

Fig. 9-20. Boring an internal contour on an engine lathe equipped with a tracing attachment.

be taken on the "back side" of the bore which is easier to see from the operator's position in front of the lathe. This should not be done on lathes having a threaded spindle nose because the cutting force will tend to unscrew the chuck.

Cutting Off

The objective of the cut-off operation is to sever the end of the workpiece which is protruding from the chuck. Often, this is a finished or semi-finished workpiece that has been machined from a bar of stock which is held in a chuck such as shown in Fig. 9-21. A cut-off tool, Fig. 9-22, is used to perform the cut-off operation. The end-cutting edge is ground at a 3-degree angle in order to cut off the part without a burr. The burr remains on the stock that is held in the chuck and can be removed by feeding the tool a short distance beyond the point where the stock is separated. The width of the cut-off blade is dependent upon the size of the cut-off tool. It should be kept as small as possible. Although both larger and smaller widths are used, the average width of the cut-off blade is from

Courtesy of Cincinnati Milacron

Fig. 9-21. Cutting off finished part in an engine lathe.

⅛ to ¼ inch. The blade has a back taper of one degree on each side in order to provide clearance for the tool with respect to the sides of the groove made by this tool. The side relief angles are two degrees and the end relief angle is ten degrees. The face of the tool is made flat, although a very large radius (equal to the radius of the grinding wheel used to prepare this surface) is sometimes ground on the face. This provides a very small back-rake angle.

Fig. 9-22. Design of a cut-off tool.

The cutting speed used to cut off stock can usually be as fast as recommended for turning. Occasionally, a very deep cut must be taken

to cut off a part, in which case the cutting speed should be reduced. This is done more in the nature of a safety measure than to decrease the cutting load on the tool. The cutting load, or cutting force, on the tool will not be reduced by decreasing the cutting speed, but by reducing the feed. For this reason, a relatively light feed should always be used to perform the cut-off operation. For very small cut-off tools a feed rate as low as .0005 inch per revolution is sometimes used. Large cut-off tools can operate satisfactorily using a feed of .010 to .012 inch per revolution. For average conditions a feed rate of .002 to .006 inch per revolution is recommended. It is best to use the automatic-power across feed to perform this operation since this feed is more uniform than the hand feed. It should be pointed out that the *actual* feed rate of the cross slide of most engine lathes is less than the *longitudinal* feed rate corresponding to the same setting of the feed-thread gear box. The setting of the feed-thread gear box will correspond to the actual longitudinal feed; however, the cross-slide feed will be less. Since the ratio of the longitudinal and cross feeds depends upon the design of the lathe, no general rule for calculating the actual cross feed rate can be given. This must be determined experimentally for each lathe.

QUESTIONS

9-1. How does a three-jaw universal chuck operate?

9-2. Explain the procedure for chucking a round part accurately, in a four-jaw independent chuck.

9-3. How does a spring collet-chuck clamp the workpiece?

9-4. Why should rough and inaccurate workpieces not be held in a collet chuck?

9-5. What is meant by facing?

9-6. What is the first step in drilling a hole in an engine lathe?

9-7. What is the danger in drilling large-diameter holes with taper shank drills that are held in the taper of the tailstock spindle?

9-8. When can a large drill be held in the taper of the tailstock spindle?

9-9. Explain how a drill holder is used to drill a hole in an engine lathe.

9-10. How can the load on the machine be reduced when drilling large holes in an engine lathe?

9-11. Why should a floating reamer-holder be used when reaming a hole in an engine lathe?

9-12. What is the objective of boring a hole with a lathe?

9-13. Why should the lead angle and the nose radius of a boring tool be kept small?

9-14. What is the objective of rough boring a hole?

9-15. How is the eccentricity of a hole corrected with a boring tool?

9-16. Describe the procedure for boring a hole to an exact size.

9-17. How is a part finished in a lathe cut off?

CHAPTER **10**

Taper and Angle Turning

Twist drills, end mills, arbors, lathe centers, other small tools, and machine parts have tapered shanks which fit into spindles, sockets, or adapters having a matching taper. Heavy-duty power generating and marine components are often designed to have tapered fits. The taper provides an accurate means of location for the mating parts. Many tapers are also designed to hold the two members together. They are called self-holding tapers in contrast to self-releasing tapers which are designed to locate, but not to hold. Tapers have been standardized in order to be interchangeable. Most machine shops are frequently called upon to measure or gage, and to machine tapers. External tapers are machined either by turning in a lathe or by grinding. Internal tapers are machined by boring in a lathe, reaming, and by internal grinding. This chapter will concern itself only with the measurement of tapers, and with cutting tapers and other conical surfaces on a lathe.

Standard Tapers

The exact dimensions and specifications for standard tapers are tabulated in engineering handbooks such as "Machinery's Handbook." The objective here is to introduce the standard tapers in common usage in the United States.

Morse Taper. This is a self-holding taper. Taper-shank twist drills are made exclusively with the Morse taper. It is also used extensively on the shanks of reamers, end mills, countersinks, counter bores, spot facers, and lathe centers. It is used on drill-press spindles and lathe spindles. Each size of Morse taper is different, although in most cases it is approximately $5/8$ inch per foot.

Brown & Sharpe Taper. This taper is approximately $1/2$ inch per foot for all sizes except No. 10 which is .5161 inch per foot. Brown & Sharpe tapers are used on the shanks of end mills and reamers, collets, milling machine spindles, and grinding-machine spindles.

Jarno Taper. The Jarno taper has been used on some machine tools. This system is based on some very simple formulas that make all the dimensions very easy to calculate. The taper per foot for all Jarno tapers is .600 inch. The following formulas are used to calculate the other dimensions:

$$\text{Diameter at large end} = \frac{\text{Jarno taper number}}{8}$$

$$\text{Diameter at small end} = \frac{\text{Jarno taper number}}{10}$$

$$\text{Length of taper} = \frac{\text{Jarno taper number}}{2}$$

Thus, a No. 7 Jarno taper has the large diameter equal to 7/8 inch, the small diameter equal to 7/10 or .700 inch, and the length is equal to 3½ inches.

Milling-Machine Taper. This taper is 3½ inches per foot in all sizes, in order to assure easy release from the milling-machine spindles of arbors, adapters, and other tools. It was adopted as a taper by the milling-machine manufacturers of the National Machine Tool Builders Association, in 1927. It has now become an American standard and is used on almost all of the milling-machine spindles built in the United States.

American Standard Machine Tapers. This standard includes a self-holding series, and a steep taper, or self-releasing, series. The self-holding tapers are, to a large extent, an adaption of the Morse and Brown & Sharpe tapers. The steep taper series corresponds closely to the Milling-Machine Taper.

There are other standard tapers, such as for American Standard Taper Pins and for the Jacobs Taper that is used for chucks and on spindles of portable air and electric tools. The wide application of all of these standards has contributed greatly to the growth of metal manufacturing.

Taper Calculations

The following definitions are used in dealing with tapered surfaces:

T = Taper; the difference in size between the large and the small diameter.

T_{pi} = Taper per inch; the change in size of the taper in a one-inch length.

T_{pf} = Taper per foot; the change in size of the taper in a one-foot length.

D = Large diameter of the taper

d = Small diameter of the taper

L = The length of the taper in inches.

The formulas used to calculate the components of a taper are given below. Their applications can best be illustrated by examples:

$$T = D - d \tag{10-1}$$

$$T_{pi} = \frac{D - d}{L} \tag{10-2}$$

$$T_{pi} = \frac{T_{pf}}{12} \qquad (10\text{-}3)$$

$$T_{pf} = 12 \times T_{pi} \qquad (10\text{-}4)$$

$$D = (T_{pi} \times L) + d \qquad (10\text{-}5)$$

$$d = D - (L \times T_{pi}) \qquad (10\text{-}6)$$

$$L = \frac{D - d}{T_{pi}} \qquad (10\text{-}7)$$

Example 10-1:

A taper is 4 inches long. The large diameter is 1 inch and the small diameter is ¾ inch. Calculate the taper per foot.

$$L = 4 \text{ in.}, \qquad D = 1 \text{ in.}, \qquad d = \tfrac{3}{4} \text{ in.}$$

$$T_{pi} = \frac{D - d}{L} = \frac{1 - \tfrac{3}{4}}{4} = .0625 \text{ in.} \qquad (10\text{-}2)$$

$$\begin{aligned} T_{pf} &= 12 \times T_{pi} = 12 \times .0625 \\ &= .750 \text{ in.} \end{aligned} \qquad (10\text{-}4)$$

Example 10-2:

Calculate the diameter of the large end of a taper when the small diameter is .778 inch, the taper per foot is .60235 inch, and the length of the taper is 3.250 inch.

$$L = 3.250 \text{ in.}, \qquad d = .778 \text{ in.}, \qquad T_{pf} = .60235 \text{ in.}$$

$$T_{pi} = \frac{T_{pf}}{12} = \frac{.60235}{12} = .05019 \text{ in.} \qquad (10\text{-}3)$$

$$\begin{aligned} D &= (T_{pi} \times L) + d = (.05019 \times 3.250) + .778 \\ &= .941 \text{ in.} \end{aligned} \qquad (10\text{-}6)$$

Measuring and Gaging Tapers

The method of gaging tapers is illustrated in Fig. 10-1. Taper plug and taper ring gages are simple and dependable for gaging tapered holes and tapered shafts, respectively. They simultaneously gage the diameter and the angularity of the taper.

To gage the angularity of a tapered shaft, place a light coating of Prussian blue along the entire length of the taper. If Prussian blue is not available, a light chalk can act as a substitute. The taper is inserted in the ring gage and twisted. If, after it has been removed, the Prussian blue coating has been uniformly rubbed off, the taper is cut at the correct angle. If a portion of the Prussian blue is untouched, the angle is not correct. The direction of the error can be determined by observing the

Ch. 10 TAPER AND ANGLE TURNING 223

Fig. 10-1. Step type plug and ring Go–Not Go taper gages.

area where the Prussian blue has been rubbed off. If it has been rubbed off toward the large end, the taper is too steep and if it has been rubbed off toward the small end, it is too gradual. Tapered holes are gaged for angularity in a similar manner, using a taper plug gage. In this case the Prussian blue, or chalk, is placed on the plug gage instead of on the workpiece. When gaging tapered holes with a taper plug gage these conditions, previously indicated for a shaft, are reversed. A self-holding taper that has the correct angle can be made to hold together without any difficulty.

The taper plug and ring gages are designed on the GO-NOT GO principle. For example, at Y, Fig. 10-1, the small end of the ring gage has two steps, or surfaces, at different levels which are designated Step A and Step B. When Step A is flush with the small end of the shaft, the basic diameter of 1.219 inches is, in this case, 1.469 inches from the small end. When Step B is flush, the basic diameter is 1.438 inches from the small end. Hence, this is a limit type of taper ring-gage and the tolerance for the position of the gage is represented by the height of the step which, in this case, equals $1.469 - 1.438 = .031$ inch. When the gage is flush with either step, or when the end of the work falls within the two steps, the taper is within the specified limits. If the end of the work is inside Step B, the taper is too large; if the end is outside Step A, the taper is too small. The plug gage at X, Fig. 10-1, also has a stepped surface and the hole is gaged in a manner similar to gaging the shaft with the taper ring-gage. The plug and ring gages determine if the taper being gaged meets the dimensional specifications; however, it does not actually measure the taper.

The angularity of the taper is measured most readily by using a sine bar or a sine plate. A sine bar is shown in Fig. 10-2. The arbor, or shaft, that is to be measured is held on two bench centers. The sine bar has two buttons which are similar to toolmakers buttons except that they are not adjustable. These buttons are usually either 10 inches or 5 inches apart. Their location on the sine bar is accurate to "tenths." The sine bar is positioned over the taper, as shown, and the difference in the height over the buttons, H_1 and H_2, is measured using a vernier height-gage and a dial test indi-

Fig. 10-2. Measuring the angle of the taper using a sine bar and precision bench-centers.

cator. The half angle of the taper is obtained from this measurement by applying the following formula:

$$\sin \theta = \frac{H_2 - H_1}{10} \quad \text{(For 10-inch sine bar)} \qquad (10\text{-}8a)$$

$$\sin \theta = \frac{H_2 - H_1}{5} \quad \text{(For 5-inch sine bar)} \qquad (10\text{-}8b)$$

The angle, θ, is found directly from a table of trigonometric functions by obtaining the angle corresponding to the calculated value of $\sin \theta$ in Formula 10-8. The included angle of the taper is two times θ. The taper per inch can be found by using the following formula:

$$T_{pi} = 2 \times \tan \theta \qquad (10\text{-}9)$$

Example 10-3:

Referring to Fig. 10-2, the vernier height-gage reading obtained for H_1 is 4.678 inches and for H_2 it is 6.122 inches. Calculate the included angle of the taper and the taper per foot of the arbor.

$$\sin \theta = \frac{H_2 - H_1}{10} = \frac{6.122 - 4.678}{10} = .1444 \text{ inch} \qquad (10\text{-}8a)$$

$$\theta = 8°18' \text{ (From a table of trigonometric functions)}$$

The included angle of the taper $= 2 \times \theta = 2 \times 8°18'$
$$= 16°36'$$

$$T_{pi} = 2 \times \tan \theta = 2 \times \tan 8°18' \qquad (10\text{-}9)$$
$$= 2 \times .14588 = .29176 \text{ inch}$$

$$T_{pf} = 12 \times T_{pi} = 12 \times .29176$$
$$= 3.501 \text{ inch} \qquad (10\text{-}4)$$

The half angle of the taper can also be determined by measuring over precision rolls as shown in Fig. 10-3. Two measurements, M and m, must be taken that are a distance H, apart. The distance, or height, above the precision-surface plate H, is obtained by two stacks of gage blocks of equal length. The half angle, θ, is calculated by the following formula:

$$\tan \theta = \frac{M - m}{2H} \tag{10-10}$$

The diameter D, at any given distance H, from the small end of the taper, can be determined by measuring the distance M over precision-gage rolls having a known diameter W by the following formula:

$$D = M - W\left[1 + \cotan \tfrac{1}{2}(90 - \theta)\right] \tag{10-11}$$

The diameter of the small end of the taper can be measured by substituting the measurement m, for M in Formula 10-11.

Example 10-4:

Referring to Fig. 10-3, the measurement across the rolls M is 2.2558 inches, and m is 2.0333 inches. The distance between the rolls H is 4.250 inches and the diameter of the rolls is .500 inch. Calculate the half angle θ.

$$\tan \theta = \frac{M - m}{2H} = \frac{2.2558 - 2.0333}{2 \times 4.250} = .026176$$

$$\theta = 1°30' \text{ (From table of trigonometric functions)}$$

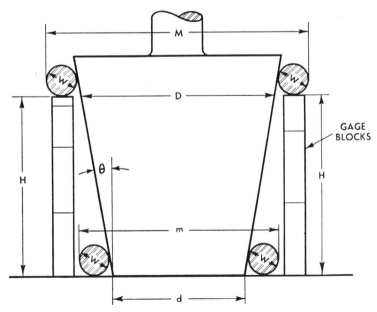

Fig. 10-3. Measuring a taper with gage rolls and precision gage-blocks.

Calculate the diameter of the taper at a distance of 4.250 inches from the small end.

$$D = M - W[1 + \cotan \tfrac{1}{2}(90 - \theta)]$$
$$= 2.2558 - .500[1 + \cotan \tfrac{1}{2}(90 - 1°30')]$$
$$= 2.2558 - .500[1 + \cotan 44°15']$$
$$= 2.2558 - .500[1 + 1.0265]$$
$$= 1.2426 \text{ inches}$$

Calculate the diameter at the small end.

$$d = m - W[1 + \cotan \tfrac{1}{2}(90 - \theta)]$$
$$= 2.0333 - .5[1 + \cotan \tfrac{1}{2}(90 - 1°30')]$$
$$= 1.0201 \text{ inches}$$

Taper Turning with a Taper Attachment

The taper attachment is a guide bar that imparts a uniform motion to the cross slide as the carriage is moved longitudinally. The rate or extent of the cross-feed movement for a given longitudinal movement determines the angle of the taper. The angle of the taper on the work corresponds to the angle at which the taper attachment is set. This can be varied and provided on the taper attachment are scales that can be read in terms of degrees and taper per foot. The taper attachment does not interfere with straight or cylindrical turning and boring. It is a permanent part of the lathe that is always ready to be used by making a few simple adjustments. A taper part is turned practically the same as a cylindrical part; that is, the power longitudinal feed is used and, as the carriage moves along the bed, the cross slide and cutting tool are gradually moved in or out by the taper attachment. The depth of cut of the cutting tool is set by using the compound-rest slide, or, if the taper attachment is a telescoping type, the cross slide can be used.

A taper attachment and a setup for turning tapers is shown in Fig. 10-4. The bar, A, is set either to the required taper per foot or to a given angle, in degrees, as determined by the graduations on the end of the bar. The nuts, D, are used to clamp the bar securely in the required position. Arm C, is clamped to the lathe bed by tightening nuts F, so that the bar, A, is held stationary. Nut G, is loosened to permit draw-bar H, to slide freely through arm C. When the attachment is not in use, nut G, is tightened so that the auxiliary draw-bar H, will insure a smooth, easy movement of arm C, along the bed without any tendency to bind. The taper bar A, may be adjusted to the exact position required, by means of the knurled knob, J. This attachment has a telescoping type of cross-feed screw permitting the cross slide to be used to adjust the tool to the depth of cut. It has a maximum capacity of 6 inches taper per foot. This exceptionally steep capacity is due to the incorporation of antifriction bearings, in order to reduce the frictional resistance.

The taper attachment can be accurately adjusted by placing a dial test indicator in the toolpost and placing the contact point against the tail-

Ch. 10　　　　　TAPER AND ANGLE TURNING　　　　　227

Courtesy of The American Tool Works

Fig. 10-4. Details of an engine-lathe taper attachment.

stock spindle, as shown in Fig. 10-5. The taper per inch of the taper to be cut is calculated and divided by two. This is the distance that the cross slide should move when the carriage is moved exactly one inch, longitudinally. The taper attachment is then adjusted until the dial indicator

Fig. 10-5. Method of setting the taper attachment. The same setup can be used to offset the tailstock.

reads the number of thousandths of an inch calculated above, when the carriage is moved exactly one inch. Care must be taken to eliminate all of the lost motion in the taper attachment before starting the dial indicator reading. This is done by moving the carriage toward the headstock until the dial indicator hand starts to turn. The one-inch, longitudinal movement of the carriage is started from this position. It must be emphasized, however, that even this method does not guarantee that the taper produced will be absolutely accurate. Accuracy is obtained by measuring or gaging the taper that has been cut and making fine adjustments of the taper attachment until the angle of the taper per foot is correct.

The point, or nose, of the cutting tool must be set at the same height as the center, or axis, of the work when turning tapers. The importance of this will be apparent by referring to Fig. 10-6. To turn the taper shown, the tool, T, would be moved back a distance X, while traversing the length, l. In order to illustrate the principle, assume that the tool could be placed as high as point a, the setting of the taper attachment

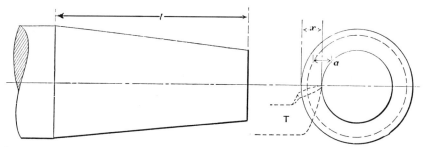

Fig. 10-6. The nose of the cutting tool must be in the same horizontal plane as the axis of the work, for taper turning.

remaining as before. When in this position, the tool would again move back a distance X, while traversing a distance l; however, in this case the large diameter would be undersize as shown by the dotted line if the diameters of the small end were the same in each case. Of course, if the tool were only slightly above or below the center, the resulting error would also be small. The tool can be set on center by comparing the height of the point, or nose, with one of the lathe centers, before placing the work in the lathe. This precaution must be observed when turning or boring any taper or angle, regardless of the method used.

With the work set up, as shown in Fig. 10-4, the entire length of the taper is cut until the taper ring-gage slips over the work far enough to allow the angle of the taper to be gaged, yet still leaving enough metal for one or more finish cuts to be taken. Several cuts are usually required and the last cut before using the taper should be a light one so that the resulting taper will accurately reflect the setting of the taper attachment. If necessary, the taper attachment is adjusted to correct any error in the angle of the taper as determined by the taper ring-gage and then another

light cut is taken along the entire length of the taper. When the angle of the taper is correct, one or more finish cuts are taken until the taper is completed. If a taper ring-gage is not available, a drill press sleeve, lathe headstock-spindle adapter, or a part having the correct taper may be used in its place. It is best, however, to use a taper gage even if one must specially be made for the job at hand.

Taper Turning—Offset-Center Method

The offset-center method of turning tapers, shown in Fig. 10-7 can be used only to turn external tapers with the work held between the head-

Fig. 10-7. Taper turning by offset-center method on lathe not equipped with a taper attachment.

stock and the tailstock centers. When both of these centers are in line, the movement of the tool is parallel to the axis of the work and, consequently, a cylindrical surface is produced. If, however, the tailstock center is set out of alignment, as shown in Fig. 10-7, the work will then be turned, tapering as the tool is traversed from a to b, because the axis x-x is at an angle with the movement of the tool. The amount of taper, or difference between the diameters at the ends for a given length, will depend upon how much the center h_1, is set over from the central position.

The amount that the tailstock center should be set over for turning a given taper can be calculated; however, the calculated distance is only a close approximation due to the effect of the centers inside the center holes

of the workpiece. Because of differences in the depth of the center holes and the length of each individual workpiece, each workpiece must be gaged to determine the angle of the taper before it is cut to the finish size. This is not necessary when using a taper attachment which will reproduce the same taper after it has been set to the correct setting. The estimated amount of offset of the center for turning a taper by the offset-center method can be calculated by the following formula:

$$\text{Offset} = \frac{T_{pi} \times L_w}{2} \qquad (10\text{-}12)$$

where: T_{pi} = Taper per inch
L_w = Length of the workpiece

Example 10-5:

A taper plug that is 7½ inches long is to be turned in a lathe by the offset-center method. The taper is 4 inches long and the taper per foot is .750 inch. Determine the tailstock offset.

$$T_{pi} = \frac{T_{pf}}{12} = \frac{.750}{12} = .0625 \text{ inch.}$$

$$\text{Offset} = \frac{T_{pi} \times L_w}{2} = \frac{.0625 \times 7.500}{2} = .234 \text{ inch}$$

To adjust the tailstock, place a dial test indicator against the tailstock spindle as was shown in Fig. 10-5. Set both the dial test indicator and the micrometer cross-feed dial to read zero. Using the graduations on the micrometer cross-feed dial to determine the distance, move the carriage back the required distance plus one turn of the cross-feed screw in order to take up the back lash of the screw. The cross slide is then moved forward one turn, plus the distance required to come to the micrometer cross-feed-dial reading. The tailstock is then moved by loosening the tailstock clamp-nut and turning the set-over screw until the tailstock spindle again comes against the dial test indicator and makes it read zero. The tailstock is then clamped in place. The procedure for turning the taper *after* the lathe has been set up, is essentially the same as is used when taper turning with a taper attachment.

Angle Turning—The Compound Rest

Steep tapers and angles can be turned by positioning the compound rest at the desired angle and feeding the cutting tool with the compound-rest slide. An example of this method is illustrated in Fig. 10-8. The correct position of the compound rest is obtained by using the angle graduations on the base of the compound rest. The compound rest can be positioned more accurately by using a vernier bevel-protractor and a dial test indicator which is attached to the compound-rest slide. The vernier bevel-protractor is held against any convenient surface on the lathe or

Fig. 10-8. Method of turning angles and steep tapers with the compound rest.

Courtesy of the Clausing Division of the Atlas Press Company

Fig. 10-9. Turning an angle on a V-belt sheave with the compound rest.

on the work and the contact point of the dial test indicator is moved along the blade of the protractor. A convenient reference surface for the protractor would be the face of a faceplate or the tailstock spindle. Very accurate settings of the compound rest are best made by taking a trial cut on the workpiece and measuring the resulting angle. The compound rest is adjusted until the angle being cut is correct. Figure 10-9 shows the angle of a V-belt sheave being turned, by using the compound rest to feed the tool. The sheave is held on a mandrel which is mounted between centers.

Boring Internal Tapers and Angles

Internal tapers can be bored by using the taper attachment as shown in Fig. 10-10. The taper attachment guide-bar A, is set to the required taper as shown by the graduations. The guide bar, locked into position by screws D, and arm C is clamped to the lathe bed by the screws shown. Lever E, is tightened and the attachment is ready to use. The point, or nose, of the boring tool must be set at the height of the lathe center. This can be done by comparing the height of the tailstock center (not shown in Fig. 10-10) with the nose of the boring tool using a surface gage, and taking the flat surfaces on the bed of the lathe as a reference surface. The procedure for boring the hole is similar to the procedure for turning a taper. The hole is bored undersize until the tapered plug-gage can be inserted to measure the angle of the taper. Following a light cut, the angle of the taper is measured and the taper attachment is adjusted to

Fig. 10-10. Use of a taper attachment to bore a taper hole in the end of a spindle.

Ch. 10 TAPER AND ANGLE TURNING 233

Courtesy of the Midwest Industrial Management Association

Fig. 10-11. Taper reaming in an engine lathe.

make the necessary small correction. The taper is then bored to size. Internal tapers can usually be finished more rapidly and more accurately with a taper reamer as shown in Fig. 10-11. The taper in the hole is first rough-bored; this is followed by reaming with a roughing reamer and the final finish is obtained by reaming with a finishing reamer. Some-

Fig. 10-12. Boring tapered holes by using compound rest set to the angle required.

times, the rough taper-boring operation is omitted and the tapered hole is made by reaming with the roughing and finishing reamers.

The tapered hole is reamed by placing a tap wrench on the square shank of the reamer and inserting the reamer into the bore. The tailstock center is inserted in the center-drilled hole in the end of the reamer. The headstock gears are disengaged to allow the spindle to be rotated by hand, without too much effort. The spindle is then rotated, manually, by turning the chuck, while at the same time, the reamer is slowly fed into the work by moving the tailstock spindle with the handwheel. A good grade of cutting oil should be used in the performance of this operation.

The compound rest can also be used to bore internal tapers and angles. An angle is being bored inside of a large diameter at A, Fig. 10-12. The diameter is large enough to allow the entire compound rest to enter the bore. At B, Fig. 10-12, an angle is being bored in a smaller hole using the compound rest. In this case a boring bar is used in order to get into the bore. Note how the boring bar is mounted on the compound rest.

QUESTIONS

10-1. Name two applications for self-holding tapers and two applications for self-releasing tapers.

10-2. Name two standard self-holding taper systems.

10-3. The large diameter of a taper is 4 inches, the small diameter is 2½ inches, and the length of the taper is 4 inches. Calculate the taper per inch.

10-4. The length of a taper is 5 inches, the small diameter is 1 inch, and the taper per inch is .050 inch. Calculate the diameter of the large end of the taper.

10-5. The large diameter of a taper is .750 inch, the small diameter is .700 inch, and the taper per foot is .250 inch. Calculate the length of the taper.

10-6. How is the GO-NOT GO ring gage used to measure the angle and the size of a taper?

10-7. (Refer to Fig. 10-2.) Using a vernier height-gage and a dial test indicator, the height H_1 was measured to be 3.752 inches and the height H_2 was 4.062 inches. Calculate: a. The included angle of the taper; b. The taper per inch; and c. The taper per foot.

10-8. (Refer to Fig. 10-2.) The measurement over the buttons showed that H_1 was 6.752 inches and H_2 was 6.250 inches. Calculate: a. The included angle of the taper; b. The taper per inch; and c. The taper per foot.

10-9. (Refer to Fig. 10-3.) The taper is measured using measuring rolls that have a diameter W equal to .500 inch. The height of the gage blocks, H, is 3.250 inches. The measurement over the rolls, M, is 1.938 inches and the measurement, m, is 1.787 inches. Calculate:

Ch. 10 TAPER AND ANGLE TURNING 235

a. The half-angle of the taper, a; b. The diameter, d, at the small end of the taper; and c. The diameter, D, at the large end of the taper.

10-10. How is the taper attachment accurately adjusted to obtain the desired taper per foot on the workpiece?

10-11. Why must the nose, or point, of the cutting tool be accurately positioned to the height of the center, or axis, of the work when turning tapers?

10-12. Describe the procedure for turning a taper with a taper attachment.

10-13. What are the limitations of turning a taper by the offset-center method?

10-14. A No. 3 Morse taper (.60235 taper per foot) is to be turned using the offset-center method. The length of the workpiece is 8 inches and the length of the taper is $3\tfrac{7}{8}$ inches. Calculate the distance that the tailstock must be offset to cut this taper.

10-15. A shaft having a total length of 10 inches is tapered at one end. The length of the taper is four inches, the small diameter is $1\tfrac{3}{8}$ inches, and the large diameter is $1\tfrac{5}{8}$ inches. Calculate the distance that the tailstock must be offset to cut this taper.

10-16. Describe the procedure used to offset the tailstock in preparation for turning a taper by the offset-center method.

10-17. When should the compound rest be used to turn a taper or an angle?

10-18. Why must the nose of the boring tool be set at the height of the lathe centers when boring a taper?

10-19. Describe the methods of finishing an internal taper on an engine lathe.

10-20. How are internal tapers cut using the compound rest?

CHAPTER **11**

Faceplate Work

As a result of their unusual shape, some workpieces cannot be conveniently held in a chuck or between the lathe centers, and faceplates are used to hold them. Often, considerable thought and planning must be used in order to determine the best way to set up the workpiece on the faceplate. Care and skill must be exercised to make certain that the part is correctly aligned on the faceplate so that the surfaces that are to be machined will be in the desired location. Faceplate work is so diversified that it can best be learned by a careful study of some typical examples. In this chapter the examples and discussions are closely related to the illustrations. The application of faceplates in obtaining an accurate hole location will be treated in detail in Chapter 15.

Faceplate Reference Surfaces

The faceplate is a reference surface that is perpendicular to the axis of rotation of the lathe. Work can be attached directly to the faceplate with the assurance that it is resting on a surface that is at right angles to that axis. Reasonable care must be exercised to prevent the faceplate from being damaged, thereby destroying its accuracy. Sometimes a faceplate does not run true on the lathe spindle. In this event a facing cut should be taken with the faceplate mounted on the spindle nose, as shown in Fig. 11-1. New faceplates should always have a truing cut taken across the face before they are used. The facing cut should be taken with a broad-nose, cast-iron finishing tool such as is shown in Fig. 11-1. The spindle speed of the lathe is determined by the outside diameter of the faceplate and it should be based on a cutting speed of about 50 fpm. The depth of cut should be just enough to "clean up" the surface of the faceplate. The compound-rest slide can be used to set the cutting tool to the depth of cut; to do this it is positioned to move parallel to the lathe axis. The feed rate for taking the broad-nose, finishing cut should be approximately .9 times the width of the face of the tool. Care must be exercised in positioning the cutting edge of the tool parallel to the cross-feed movement, otherwise an undesirable groove will be cut on the faceplate. This entire operation should be done in one or two shallow cuts (.001 to .003-inch deep) across the faceplate.

An angle plate, Fig. 11-5, can be attached to the faceplate to provide a reference surface that is parallel to the lathe axis. The angle plate must

Fig. 11-1. Truing the surface of a faceplate by taking a facing cut with a broad-nose, cast-iron finishing tool.

be square (this should be verified with a precision square); if it is not, it should be machined square on a planer or a shaper. Sometimes angle plates having angles other than 90 degrees are used. In either case the workpiece is attached to the angle plate.

Faceplate Jaws

Faceplate jaws for clamping workpieces are bolted onto the faceplate to provide movable jaws similar to a four-jaw independent chuck. Faceplate jaws are shown in Figs. 11-2 and 8-2. They are used primarily on larger faceplates that are mounted on large lathes in place of four-jaw independent chucks which, in the larger sizes, would add an unnecessarily heavy weight to the nose of the spindle. A chuck would also place the weight of the workpiece further away from the support of the spindle

Courtesy of The R. K. LeBlond Machine Tool Company

Fig. 11-2. Faceplate jaws used to hold workpieces for facing operation.

bearings, which is undesirable. This is avoided when using faceplate jaws which are just as effective as a chuck in holding the workpiece. Faceplate jaws can be used to hold symmetrical parts, as shown in Fig. 11-2. They can also be used to hold irregularly shaped parts, as the different jaws can be located at any distance from the center of the faceplate.

Faceplate Setups

The illustrations in this chapter are used as examples of the application of faceplate work on engine lathes. Although the same setup is not likely to be encountered, there is frequently a similarity between the jobs that are illustrated in this chapter and those that are encountered elsewhere. For this reason the examples shown should be carefully analyzed.

Figure 11-3 shows a rectangular-shape casting with a round boss, or projection, the end of which is to be turned parallel with the finished back face of the casting. A cored hole through the center of the boss is also to be bored true. This is a typical example of a part that can be conveniently machined by bolting it directly to the faceplate of the lathe. To perform this operation in the lathe, clamp the finished surface of the casting directly against the faceplate by bolts and clamps, a, b, c, and d, as shown. The work would then be turned in the same way as if it were held in a chuck. By holding the casting in this way, the face, e, will

be finished parallel with the back surface, because the latter is clamped directly against the true-running surface of the faceplate. If a casting of this shape were small enough it could also be held in the jaws of a four-jaw independent chuck, but if the surface, *e*, needs to be exactly parallel with the back face, or if the bore must be exactly perpendicular to this surface, it is better to clamp the work to the faceplate.

The proper way to clamp a piece to the faceplate depends largely upon its shape, the location of the surface to be machined, and the location of any previously machined surfaces. Whenever possible, previously machined surfaces should be clamped against the faceplate as they provide reference surfaces from which the work can be aligned. When the work is bolted to an angle plate instead of directly to the faceplate, the previously machined surfaces should be placed against the angle plate. If a hole is to be bored through a workpiece which is larger than the bore in the spindle nose, the workpiece should be clamped against parallel bars, and not directly against the faceplate, to provide clearance for the tool when it reaches the inner end of the hole and to prevent it from cutting the faceplate. The parallels should be located directly underneath the clamps so that the clamping forces will not bend the workpiece.

When deep roughing cuts have to be taken, especially on large diameters, it is well to bolt a piece to the faceplate and against one side of the workpiece, as at D, Fig. 11-3, to act as a driver and to prevent the work from shifting. A faceplate driver is always placed to the rear, as determined by the direction of rotation, since the work tends to shift backwards when a cut is being taken. If the surface which is clamped against the faceplate is finished, the work will be less likely to shift when a piece of paper is placed between it and the faceplate.

Sometimes, an assembly of parts is mounted on a faceplate for turning, boring, and facing operations. When machined in this manner, practically perfect alignment can be obtained between all of the parts in the assembly. Figure 11-4 illustrates boring and facing operations being performed on an assembly of parts which is held on the faceplate.

Fig. 11-3. Casting clamped to faceplate for turning and boring.

240 FACEPLATE WORK Ch. 11

Courtesy of The R. K. LeBlond Machine Tool Company

Fig. 11-4. Boring and facing operations performed on an assembly of parts that are clamped directly to a faceplate.

An angle plate is used in conjunction with a faceplate, for holding elbows or other parts that cannot be clamped against a vertical surface. Figure 11-5 illustrates how an angle plate is used to hold a cast-iron elbow, E, the two flanges of which are to be faced true and square with each other. The shape of the casting is such that a flange cannot be held on the faceplate while the other flange is being machined. The angle plate, P, provides the reference surface against which a flange can be bolted

Fig. 11-5. Cast elbow held on angle plate attached to faceplate.

while the other flange is machined at right angles to it. When setting up an angle plate for this operation, the distance from its work-holding side to the center of the faceplate is made equal to the distance, d, between the center of one flange and the face of the other.

As the angle plate and the work are almost entirely on one side of the faceplate, a weight, W, is attached to the opposite side for counterbalancing. Very often, weights are also needed to counterbalance offset parts that are bolted directly to the faceplate. The necessity for counterbalancing depends somewhat upon the spindle speed used for turning. If the surface to be machined is small in diameter so that the lathe can be run quite fast, any unbalanced part should always be counterbalanced. Likewise, if the counterbalancing is not perfect, which is often the case, the spindle speed must be reduced to avoid excessive vibrations. The faceplate and the parts attached to it can be counterbalanced so that no vibrations result from this assembly. It is important, however, to understand that even when balanced, heavy workpieces will place a very severe load on the clamps due to the action of centrifugal force. Thus, the speed of the spindle should be limited when heavy, unbalanced masses are encountered, even if these masses are balanced by counterweights. Balance of the faceplate assembly is achieved only when the force tending to move the counterweights is equal and opposite to the force tending to move the workpiece.

Figure 11-6 illustrates a part being held and bored while clamped to an angle plate that is attached to the faceplate. This is a convenient method of boring and facing parts that cannot otherwise be clamped and located

Courtesy of the Clausing Division of the Atlas Press Company

Fig. 11-6. Boring a part that is clamped to an angle plate attached to the faceplate.

in the lathe. The base of this part has been previously machined and is used as a locating surface for holding the part on the angle plate. The setup for this and for similar operations is usually made in two stages. First, the angle plate is set up so that the perpendicular distance from the clamping surface of the angle plate to the axis of the lathe is equal to the distance from the reference surface on the work to the center of the bored hole. Second, the workpiece must be located along the angle plate so that the axis of the hole to be machined (or surface to be turned) coincides with the axis of the lathe.

The procedure for setting up the workpiece is illustrated in Figs. 11-7 and 11-8. A layout should be made on the workpiece before starting to make the setup on the lathe. The layout lines on the workpiece can be seen in Fig. 11-8. These layout lines define two imaginary planes which intersect at the axis of the hole. The setup is started by clamping the

Fig. 11-7. Setting up the angle plate on the faceplate of the lathe.

Ch. 11　　　　　　　FACEPLATE WORK　　　　　　　　**243**

Fig. 11-8. Aligning the workpiece on the angle plate.

angle plate to the faceplate in its approximate position. The angle plate is clamped firmly, yet loose enough so that it can be moved about on the faceplate by tapping with a hammer against a piece of soft metal which is held against the angle plate. The lathe spindle is turned to bring the clamping surface of the angle plate into a horizontal position, as shown in Fig. 11-7, and precision parallel bars are then used to form a bridge over the ways of the bed. With the surface gage placed on the parallels, the scriber point of the surface gage is set at the height of the lathe axis by making it correspond to the point of a lathe center held in the tailstock spindle. A combination square is placed on the clamping surface of the angle plate, as shown. The distance that this surface falls below the lathe axis can then be read directly, by bringing the scriber point against the graduations on the combination square rule. This height must be checked on both sides of the angle plate, as shown. The readings in both of these positions must be equal otherwise correct distance from the lathe axis to the angle plate is not obtainable. The angle

Courtesy of South Bend Lathe

Fig. 11-9. Boring a bracket mounted to an angle plate which is held on a faceplate.

Courtesy of Lassy Tools Inc.

Fig. 11-10. V-block angle plate attached to faceplate and used to hold T-shaped part.

plate is then adjusted until this distance is correct, after which it is securely clamped to the surface plate. The workpiece is then clamped to the angle plate in the approximate position in which it will be machined. The lathe spindle is turned until the work-holding surface of the angle plate is in a vertical position, as shown in Fig. 11-8, using a combination square to check this position. The scriber point of the surface gage is positioned at the height of the lathe centers as before. The location of the workpiece is then adjusted on the angle plate until the layout line, abc, is at the height of the surface-gage scriber point at a, b, and c. When in this position the setup is completed by tightening the clamps that hold the work to the angle plate. It would have been possible to locate the position of the angle plate on the faceplate by aligning the layout line that is perpendicular to abc with the surface gage; however, the method of doing this, described previously, is usually more accurate and, therefore, preferred.

Another setup for boring a part that is held on an angle plate is shown in Fig. 11-9. This angle plate is similar to those shown in Figs. 11-7 and 11-8. It is different, however, from those shown in Figs. 11-5 and 11-6

Courtesy of Lassy Tools Inc.

Fig. 11-11. V-block angle plate with special jaws used to hold part for drilling, facing, and turning.

in respect to the location of the work-holding surface. The angle plate in Fig. 11-9 has the work-holding surface on the inside face of the angle plate, while the angle plates in Figs. 11-5 and 11-6 have the work-holding surface on the outside face. In lathe work, having this surface on the inside face has the advantage of allowing larger workpieces to be held.

A very convenient fixture for faceplate work is the V-block angle plate shown in Figs. 11-10 and 11-11. This angle plate has two surfaces attached to the body, that are 90 degrees to each other—forming a V-locating surface. The fixed V-locator in the body is perpendicular to the face of the angle plate. An adjustable clamp, which also has two surfaces in the form of a V, is used to secure the part. The position of this fixture on the angle plate may be adjusted to accommodate the size and form of different workpieces. In Fig. 11-10, the clamp is shown in the open position. The workpiece in this illustration is a T-shaped part that would be very difficult to hold in any other manner. Note that the clamp is actuated by a cam. Special V-shaped jaws are used to hold the part in the V-block angle plate in Fig. 11-11. Two counterweights have been attached

Courtesy of South Bend Lathe

Fig. 11-12. Boring a jig plate with a faceplate chuck.

Ch. 11　　　　　　FACEPLATE WORK　　　　　　247

to this angle plate in order to allow higher spindle speeds to be used in drilling this part. The part in Fig. 11-11 is being drilled, turned, and faced in one setup.

A faceplate chuck made from a standard faceplate is shown in Fig. 11-12. Radial slots are on the face of the faceplate in which the workholding clamps are placed. A series of holes is drilled and tapped in the center of each slot in which the socket-head cap screws that hold the clamps are screwed. Thus, the clamps can be placed in a number of positions on the faceplate. These clamps do not hold the sides of the workpiece, which is a part of a drill jig; they clamp it against the faceplate.

QUESTIONS

11-1. How is the surface of a faceplate made perpendicular to the axis of the lathe?

11-2. What are the faceplate jaws and why are they used?

11-3. What are the two stages in setting up a workpiece that is held on an angle plate which, in turn, is attached to the faceplate?

11-4. In locating the position of the angle plate on the faceplate, why is it necessary to check the position of the angle plate in two places, as shown in Fig. 11-7?

11-5. Referring to Fig. 11-8, how are the layout lines on the workpiece used to position it on the angle plate?

11-6. Why are counterbalance weights sometimes used for faceplate work?

11-7. Describe the total effect of an unbalanced weight that is bolted to a faceplate.

11-8. Name the factors that must be considered in selecting the spindle speed of the lathe when doing faceplate work.

11-9. What is the advantage of machining an assembly of parts?

11-10. What is the V-block angle plate and how is it used?

CHAPTER **12**

Screw Threads and Screw-Thread Measuring

The majority of screw threads used in machine construction conform to some established standard. These screw-thread standards include the profile or cross-sectional shape of the thread, a range of screw thread diameters, and a standard number of threads per inch for each diameter. Tolerances and allowances for obtaining varying degrees of accuracy and different classes of fits are also included in modern screw-thread standardization. Special screw threads are used when there is some mechanical reason for doing so. There are two basic kinds of screw threads—locking or holding threads that are used on fasteners to hold parts together, and power-transmission threads that are used to transmit power and motion. Sometimes locking threads are used to transmit motion when the power required is low. When this is done, however, there is considerable loss in mechanical efficiency in comparison to threads designed to transmit power.

Definitions of Screw-Thread Terms

The definitions of the principal terms relating to screw threads are given below (see Fig. 12-1):

Major Diameter—The largest diameter of a straight screw thread. The term major diameter applies to both internal and external threads.

Minor Diameter—The smallest diameter of a straight screw thread. The term minor diameter applies to both external and internal threads.

Pitch—The distance from a point on a screw-thread profile to a corresponding point on the next thread profile, measured parallel to the axis. The pitch is equal to one divided by the number of threads per inch, or:

$$\text{Pitch} = \frac{1}{\text{No. Threads per inch}}$$

Pitch Diameter—The diameter of an imaginary cylinder, the surface of which would pass through the threads on all sides at such points as to make the width of the groove equal to one-half of the pitch. The pitch diameter is independent of the major and minor diameters. It is a measure of the diameter of the sloping sides of the thread profile. As such, it deter-

Ch. 12 SCREW THREADS AND SCREW-THREAD MEASURING

Fig. 12-1. Illustration showing screw-thread terms.

mines the size of the thread and how closely, or loosely, it fits onto its mating thread.

Nominal Size—The designation used to describe the size of the thread for purposes of general identification. For example, the nominal size of a ½-inch, 20-thread per inch, screw thread is ½ inch, but its actual major diameter for Class 2A limits might be from .4987 to .4906 inch.

Actual Size—The actual measured size.

Basic Size—The theoretical size from which the size limits are derived by the application of allowance and tolerance.

Design Size—The size from which the limits of size are derived by the application of tolerance.

Tolerance—The difference between the maximum and minimum limits of size, or the total permissible variation in size.

Allowance—An intentional difference between the maximum material limits of mating parts. It is the minimum clearance between freely fitting parts, or the maximum interference between parts having an interference fit.

Limits of Size—These are the maximum and the minimum sizes.

Maximum Material Limit—A maximum material limit of size is the maximum limit of size of an external dimension or the minimum limit of size of an internal dimension.

Minimum Material Limit—A minimum material limit of size is the minimum limit of size of an external dimenson or the maximum limit of size of an internal dimension.

Fit—The relationship existing between two mating parts with respect to the amount of clearance or interference between them when they are assembled.

Basic Form of Thread—The theoretical profile of the thread for a length of one pitch in the axial plane. It is the form on which the design forms of both external and internal threads are based.

Design Form of Thread—The design forms for a thread are the maximum material forms permitted for external and internal threads.

Lead—The distance that a screw thread advances axially in one turn. On a single thread screw the pitch and the lead are equal; on a double thread screw the lead is twice the pitch; on a triple thread screw the lead is three times the pitch, etc.

Angle of Thread—The angle included between the sides of the thread measured in an axial plane.

Lead Angle—This is the angle made by the helix of the thread at the pitch diameter, measured in an axial plane.

Crest—The top surface joining the two sides, or flanks, of the thread.

Root—The bottom surface joining the two adjacent sides or flanks of the thread.

Flank—The flank (or side) of a thread surface connecting crest and roots.

Truncation—The truncation of the root or crest of a thread is the distance, measured perpendicular to the axis, between a sharp V-shaped root or crest (apex) and the cylinder which bounds the root or crest.

Depth of Engagement—The depth of thread contact or overlap of two mating thread forms, measured radially.

Crest Clearance—In a thread assembly this is the distance, measured perpendicular to the axis, between the crest of a thread and the root of its mating thread.

Right-Hand Thread—A thread with the grooves cut so that a machine screw or a bolt would have to be turned in a clockwise direction, when viewed from the head end, in order to assemble into a threaded hole.

Left-Hand Thread—A thread with the grooves cut so that a machine screw or a bolt would have to be turned in a counterclockwise direction, when viewed from the head end, in order to assemble into a threaded hole.

Multiple Thread—A thread having two or more separate grooves cut adjacent to each other and winding around the screw in the manner of a helix. The two adjacent threads combine to produce the basic form of the thread.

Standard Screw-Thread Forms and Systems

The purpose of this section is to study the basic screw-thread forms and the formulas used to obtain their principal dimensions. Screw-thread systems include, in addition to the basic screw-thread forms, specifications for thread per inch corresponding to a given diameter, classes of fits, and basic dimensions for each thread series. These details are found in engineering handbooks such as *Machinery's Handbook*.

American Standard Unified-Thread Form. The former American National Form of Thread was used for many years in making most of the locking type threads produced in the United States. This form is practically the same as the present standard thread form used in the United States, which is called the American Standard Unified-Thread Form (Fig. 12-2). These two thread forms are completely interchangeable although

Ch. 12 SCREW THREADS AND SCREW-THREAD MEASURING 251

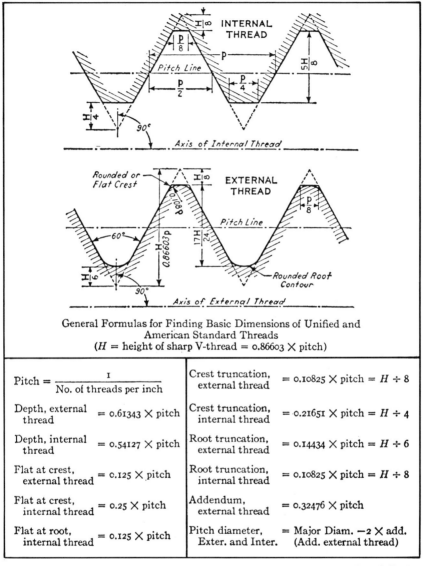

Fig. 12-2. American Standard Unified Internal and External Screw Thread Design Forms.

some differences between them do exist. The principal differences relate to the depth of the thread and the shape of the crest and root. The depth of the former American National Thread Form is .64952 × pitch while the depth of the American Standard Unified Thread Form for external threads is .61343 × pitch. The American Standard Unified Thread Form provides an option of either a round or a flat crest while the former Ameri-

252 SCREW THREADS AND SCREW-THREAD MEASURING Ch. 12

can National Thread Form has a flat crest and root. The exact form of the rounding is not specified in the American Standard which, in practice, varies according to tool wear. The width of the flat at the crest and the root is the same for both thread forms, being .125 × pitch. According to British specifications the root on the internal threads is rounded to a radius of .10825 × pitch and on the external threads, to a radius of .14434 × pitch. The preferred American practice is to have a flat crest, while the preferred British practice is to have a rounded crest. The crest of the internal thread is flat. The angle of the thread is 60 degrees.

The American Standard Unified Thread System—This is the system agreed upon by the United States, Canada, and the United Kingdom in

Table 12-1. Coarse-Thread Series, UNC and NC—Basic Dimensions

Sizes	Basic Major Diam., D	Thds. per Inch, n	Basic Pitch Diam.,a E	Minor Diameter Ext. Thds., K_s	Minor Diameter Int. Thds., K_n	Lead Angle at Basic P.D. Deg.	Lead Angle at Basic P.D. Min.	Area of Minor Diam. at $D-2h_b$	Tensile Stress Area
	Inches		Inches	Inches	Inches	Deg.	Min.	Sq. In.	Sq. In.
1 (.073)*	0.0730	64	0.0629	0.0538	0.0561	4	31	0.00218	0.00263
2 (.086)	0.0860	56	0.0744	0.0641	0.0667	4	22	0.00310	0.00370
3 (.099)*	0.0990	48	0.0855	0.0734	0.0764	4	26	0.00406	0.00487
4 (.112)	0.1120	40	0.0958	0.0813	0.0849	4	45	0.00496	0.00604
5 (.125)	0.1250	40	0.1088	0.0943	0.0979	4	11	0.00672	0.00796
6 (.138)	0.1380	32	0.1177	0.0997	0.1042	4	50	0.00745	0.00909
8 (.164)	0.1640	32	0.1437	0.1257	0.1302	3	58	0.01196	0.0140
10 (.190)	0.1900	24	0.1629	0.1389	0.1449	4	39	0.01450	0.0175
12 (.216)*	0.2160	24	0.1889	0.1649	0.1709	4	1	0.0206	0.0242
¼	0.2500	20	0.2175	0.1887	0.1959	4	11	0.0269	0.0318
⁵⁄₁₆	0.3125	18	0.2764	0.2443	0.2524	3	40	0.0454	0.0524
⅜	0.3750	16	0.3344	0.2983	0.3073	3	24	0.0678	0.0775
⁷⁄₁₆	0.4375	14	0.3911	0.3499	0.3602	3	20	0.0933	0.1063
½	0.5000	13	0.4500	0.4056	0.4167	3	7	0.1257	0.1419
⁹⁄₁₆	0.5625	12	0.5084	0.4603	0.4723	2	59	0.162	0.182
⅝	0.6250	11	0.5660	0.5135	0.5266	2	56	0.202	0.226
¾	0.7500	10	0.6850	0.6273	0.6417	2	40	0.302	0.334
⅞	0.8750	9	0.8028	0.7387	0.7547	2	31	0.419	0.462
1	1.0000	8	0.9188	0.8466	0.8647	2	29	0.551	0.606
1⅛	1.1250	7	1.0322	0.9497	0.9704	2	31	0.693	0.763
1¼	1.2500	7	1.1572	1.0747	1.0954	2	15	0.890	0.969
1⅜	1.3750	6	1.2667	1.1705	1.1946	2	24	1.054	1.155
1½	1.5000	6	1.3917	1.2955	1.3196	2	11	1.294	1.405
1¾	1.7500	5	1.6201	1.5046	1.5335	2	15	1.74	1.90
2	2.0000	4½	1.8557	1.7274	1.7594	2	11	2.30	2.50
2¼	2.2500	4½	2.1057	1.9774	2.0094	1	55	3.02	3.25
2½	2.5000	4	2.3376	2.1933	2.2294	1	57	3.72	4.00
2¾	2.7500	4	2.5876	2.4433	2.4794	1	46	4.62	4.93
3	3.0000	4	2.8376	2.6933	2.7294	1	36	5.62	5.97
3¼	3.2500	4	3.0876	2.9433	2.9794	1	29	6.72	7.10
3½	3.5000	4	3.3376	3.1933	3.2294	1	22	7.92	8.33
3¾	3.7500	4	3.5876	3.4433	3.4794	1	16	9.21	9.66
4	4.0000	4	3.8376	3.6933	3.7294	1	11	10.61	11.08

* Secondary sizes.
a British: Effective Diameter.

Ch. 12 SCREW THREADS AND SCREW-THREAD MEASURING

order to obtain screw thread interchangeability among these nations. The transition of industry from the use of the former American National thread system to the use of the unified threads is practically completed.

American Standard Unified Thread Series—Unified threads consist of a Coarse-Thread Series, Table 12-1; a Fine-Thread Series; and an Extra-Fine Thread Series; Table 12-2. There are also 4-, 6-, 8-, 12-, 16-, 20-,

Table 12-2. Fine-Thread Series, UNF and NF—Basic Dimensions

Sizes	Basic Major Diam., D	Thds. per Inch, n	Basic Pitch Diam.,a E	Minor Diameter		Lead Angle at Basic P.D.		Area of Minor Diam. at $D-2h_b$	Tensile Stress Area
				Ext. Thds., K_s	Int. Thds., K_n	Deg.	Min.		
	Inches		Inches	Inches	Inches			Sq. In.	Sq. In.
0 (.060)	0.0600	80	0.0519	0.0447	0.0465	4	23	0.00151	0.00180
1 (.073)*	0.0730	72	0.0640	0.0560	0.0580	3	57	0.00237	0.00278
2 (.086)	0.0860	64	0.0759	0.0668	0.0691	3	45	0.00339	0.00394
3 (.099)*	0.0990	56	0.0874	0.0771	0.0797	3	43	0.00451	0.00523
4 (.112)	0.1120	48	0.0985	0.0864	0.0894	3	51	0.00566	0.00661
5 (.125)	0.1250	44	0.1102	0.0971	0.1004	3	45	0.00716	0.00830
6 (.138)	0.1380	40	0.1218	0.1073	0.1109	3	44	0.00874	0.01015
8 (.164)	0.1640	36	0.1460	0.1299	0.1339	3	28	0.01285	0.01474
10 (.190)	0.1900	32	0.1697	0.1517	0.1562	3	21	0.0175	0.0200
12 (.216)*	0.2160	28	0.1928	0.1722	0.1773	3	22	0.0226	0.0258
¼	0.2500	28	0.2268	0.2062	0.2113	2	52	0.0326	0.0364
⁵⁄₁₆	0.3125	24	0.2854	0.2614	0.2674	2	40	0.0524	0.0580
⅜	0.3750	24	0.3479	0.3239	0.3299	2	11	0.0809	0.0878
⁷⁄₁₆	0.4375	20	0.4050	0.3762	0.3834	2	15	0.1090	0.1187
½	0.5000	20	0.4675	0.4387	0.4459	1	57	0.1486	0.1599
⁹⁄₁₆	0.5625	18	0.5264	0.4943	0.5024	1	55	0.189	0.203
⅝	0.6250	18	0.5889	0.5568	0.5649	1	43	0.240	0.256
¾	0.7500	16	0.7094	0.6733	0.6823	1	36	0.351	0.373
⅞	0.8750	14	0.8286	0.7874	0.7977	1	34	0.480	0.509
1	1.0000	12	0.9459	0.8978	0.9098	1	36	0.625	0.663
1⅛	1.1250	12	1.0709	1.0228	1.0348	1	25	0.812	0.856
1¼	1.2500	12	1.1959	1.1478	1.1598	1	16	1.024	1.073
1⅜	1.3750	12	1.3209	1.2728	1.2848	1	9	1.260	1.315
1½	1.5000	12	1.4459	1.3978	1.4098	1	3	1.521	1.581

* Secondary sizes.
a British: Effective Diameter.

28-, and 32-thread series, the pitch in each of these cases being constant for all diameters. The Coarse Series is recommended for general use where conditions do not require a fine thread. The Fine, and Extra-Fine, Series are particularly useful in the automotive and aircraft industries. The Fine Series is also used where the length of engagement of the threads is short, where a smaller lead angle is desired, and where the wall thickness demands a fine pitch. The 8-thread series was originally intended for high-pressure-joint bolts and nuts but is now widely used as a substitute for the Coarse-Thread Series, for diameters larger than one inch. The 12-thread series was originally intended for boiler practice. It is now

used extensively as a continuation of the Fine-Thread Series, for diameters larger than $1\frac{1}{2}$ inches. The 16-thread series is used for adjusting collars, bearing retaining nuts, and other applications requiring a fine thread. It also provides a continuation of the Extra-Fine-Thread Series for diameters larger than $1\frac{11}{16}$ inch. The 20-, 28-, and 32-thread series have been used more or less widely in industry for various applications where the Standard Coarse, Fine, or Extra-Fine Series could not be used. They are now given recognition as Standard Unified Thread Series in a specified selection of diameters for each pitch.

Each thread series has been given a symbol for easy identification. These are given below:

UNC	Unified Coarse Series
UNF	Unified Fine Series
UNEF	Unified Extra-Fine Series
UN	Used for all constant-pitch series, which are given as follows: 4UN, 6UN, 8UN, 12UN, 16UN, 20UN, 28UN, and 32UN
UNS	Special threads designed through applications of formulas given in the Unified Standard, but not given in this text.

Thread Classes—Thread classes are used to specify the fit between mating threads and the tolerance to which the thread must be made. The detailed specification for each thread is to be found in engineering handbooks; however, it is helpful to have a general understanding of the difference between the classes because of the wide usage of the terms associated with this subject. The classes are identified by a numeral followed by a letter A and B. The numeral designates the class of fit, while the A designates an external thread and the B designates an internal thread.

Classes 1A and 1B—These classes are intended for ordnance work and other special applications where quick and easy assembly is necessary. They have a liberal allowance to permit the parts to assemble easily, even with slightly bruised or dirty threads. The tolerances specified for the 1A, or external, threads are not the same as those specified for the 1B, or internal, threads.

Classes 2A and 2B—These classes are the most commonly used for general applications including bolts, nuts, screws, and similar fasteners. This series provides for a *minimum clearance* between mating threads to permit the threads to be assembled without being excessively loose, but also without binding or seizing.

Classes 3A and 3B—These classes have closer tolerances than those provided by Classes 2A and 2B. The maximum diameter of a Class 3A, or external, thread is equal to the minimum diameter of a Class 3B, or internal, thread. Therefore, no clearance or allowance is provided for the assembly of maximum metal components and mating threads will bind or fit very tightly at assembly.

Ch. 12 SCREW THREADS AND SCREW-THREAD MEASURING

The selection of a screw-thread class depends upon the end use of the threads. In almost all cases a given class of thread will be available to provide required fit. When the requirements for screw-thread fit cannot be met by a single class, it is possible to specify a proper combination of thread classes for the components. For example, a class 2A external thread may be used with a class 1B, 2B, or 3B internal thread.

Screw Thread Designation—Standard thread designations are given on drawings and specifications in order to specify a given thread. These specifications are frequently encountered on machine drawings, and their meaning should be committed to memory. The standard method of designating a screw thread is to specify, in sequence, the nominal size, number of threads per inch, thread series symbol, and thread-class symbol. For example:

$$\tfrac{1}{4}\text{--}20 \text{ UNC --2A}$$

This designation specifies a thread having a nominal size of ¼ inch, 20 threads per inch, and that it is a Coarse-Thread Series having a class 2A tolerance for external threads. The pitch-diameter limits may also be designated, but are not required. The thread in the above example may be specified as follows:

$$\tfrac{1}{4}\text{--}20 \text{ UNC--3A}$$
$$\text{PD } 0.2175\text{--}0.2147$$

The additional designation specifies that the limits of size of the pitch diameter are .2175 to .2147 inch.

It is occasionally necessary to modify the limits of size of the major diameter of an external thread or the minor diameter of an internal thread from the limits in the standard in order to fit a special purpose, but without changing the class of the thread or the pitch diameter. In such cases the established thread designation is used which is followed by a statement of the modified diameter limits and the designation "MOD." For example:

$$\tfrac{3}{4}\text{--}10 \text{ UNC--2A}$$
$$\text{Major Diameter } 0.740\text{--}0.735 \text{ MOD}$$

American National Thread System—There are some older drawings used in machine shop practice on which threads are specified according to the American National Thread System; therefore, a brief description of thread specifications in accordance to this system will be given. The designations used are: NC, NF, N-8, N-12, and N-16 which indicate National Coarse, National Fine, 8-pitch, 12-pitch, and 16-pitch series respectively. There are three classes of fits of which Class 1 is the least exacting, Class 2 is recommended for general use, and Class 3 is the most exacting. The thread is designated by specifying in sequence, the nominal size, number of threads per inch, thread-series symbol, and thread-class symbol. For example, a ½ inch, 13 threads per inch thread having a Class 2 fit would be specified as follows:

$$\tfrac{1}{2}\text{--}13 \text{ NC--2}$$

American Standard for Unified Miniature Screw Threads—This standard introduces a new thread series that is intended for general purpose fastening screws and similar uses on watches, instruments, and other mechanisms. The basic profile of this thread is identical with the Unified, and American, basic thread form except for the basic height and depth of engagement which is .52 × pitch instead of .54127 × pitch. The included angle of the thread is 60 degrees and the thread-series symbol is UNM. There are fourteen sizes, and one class of thread which has a zero allowance on all diameters. The basic sizes and the pitches for this thread are metric, although the engineering tables give the equivalent inch measurements as well as the metric.

American Standard Microscopic Objective Threads—This thread system is based on the thread system originated by the Royal Microscopical Society of Great Britain, known as the RMS thread. This system is used for microscopic objective mountings and for other optical assemblies of microscopes and associated apparatus. The basic thread form is like the British Standard Whitworth thread having a 55-degree angle of thread and rounded threads and roots.

Interference Fit Threads—Interference-fit threads are holding threads in which the externally threaded member is larger than the internally fitted member. They are assembled by expanding the internal thread with the application of heat or by shrinking the external thread by subjection to a low temperature. At assembly they become the same size and develop a great holding power by elastic compression of the members. By custom, these threads are designated as Class 5.

American Standard Acme Screw Threads—The American Standard for Acme Screw Threads provides for three types of Acme threads, namely General Purpose, Stub, and Centralizing. General Purpose Acme threads have sufficient clearance on all diameters for free movement whereas Centralizing Acme threads have the clearance limited at the major diameters of the external and internal threads so that the threads can contact each other at these surfaces in order to maintain an approximate alignment of the thread axis, thereby preventing the thread flanks from wedging and causing the thread to bind. There is also an alternate American Standard for Centralizing Acme threads that has the clearance limited on the minor diameter instead of on the major diameter which then acts as a centralizing control. The Centralizing Acme thread has a particular advantage in supporting long, unsupported screws such as the lathe lead screw. The Stub Acme thread is used for unusual applications where, due to mechanical or metallurgical considerations, a coarse-pitch thread of shallow depth is required.

The thread form for General Purpose Acme threads and for Stub Acme threads is shown in Fig. 12-3. The angle of thread is 29°. The crest corners may, as an option, be chamfered to an angle of 45° and a maximum depth of $\frac{P}{15}$, where P is the pitch. For Standard Acme threads, the recommended combinations of pitch diameter and threads per inch are: $\frac{1}{4}$-16; $\frac{5}{16}$-14;

Ch. 12 SCREW THREADS AND SCREW-THREAD MEASURING

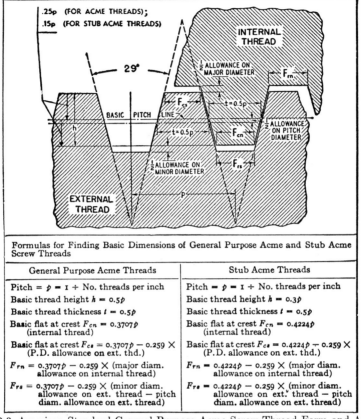

Fig. 12-3. American Standard General Purpose Acme Screw Thread Form and American Standard Stub Acme Screw Thread Form.

3/8-12; 7/16-12; 1/2-10; 5/8-8; 3/4-6; 7/8-6; 1-5; 1 1/8-5; 1 1/4-5; 1 3/8-4; 1 1/2-4; 1 3/4-4; 2-4; 2 1/4-3; 2 1/2-3; 2 3/4-3; 3-2; 3 1/2-2; 4-2; 4 1/2-2; 5-2. Other combinations can, however, be used for special applications. Three classes of General Purpose threads are provided in the standard. The Class 2G is the preferred class for most general-purpose thread assemblies. If less backlash, or end play, between the nut and the screw is required, the Class 3G or Class 4G can be used. The following examples show how Acme threads are designated on drawings and tools:

1. 1 3/4-4 Acme-2G indicates a General Purpose Class 2G Acme thread of 1 3/4 inch major diameter and 4 threads per inch, single thread, right hand. The same thread, but left hand, would be designated 1 3/4-4 Acme-2G-LH.

2. 2 7/8-0.4p-0.8L Acme-3G indicates a General Purpose Class 3G Acme thread of 2 7/8-inches major diameter, pitch .4 inch, lead .8 inch, double thread, right hand.

3. ½-20-Stub Acme indicates a ½-inch major diameter, 20 threads per inch, right hand, single thread, Standard Stub Acme Thread.

The Acme thread is extensively used in preference to the square thread because it is stronger and it may be cut more readily with a die or a tap than a square thread. The split nut on the lathe apron may be more readily engaged on an Acme thread; therefore, the Acme thread is preferred on lathe lead screws.

Square Thread—A square thread is shown in Fig. 12-4. The square thread has not been standardized and has been superseded, to a large extent, by the Acme thread. The pitch of a square thread is usually made twice the pitch of an American Standard thread of corresponding diameter. The thread groove of a square-threaded screw is .5 × pitch. The groove in the nut is made slightly larger than one-half the pitch in order to provide a clearance for the screw. There is also a 10-degree Modified Square Thread which has an angle of thread equal to 10 degrees. This thread is practically equivalent to a square thread and yet is capable of economical production.

American Standard Taper Pipe Threads—This thread has an angle of thread of 60 degrees when measured in an axial plane, and the line bisecting this angle is perpendicular to the axis. The depth of the pipe thread is greater than the American Standard Unified Thread Form and it equals .8 × pitch except for 8 threads per inch which is .788 × pitch. The tapering pipe thread has a standard taper of ¾ inch per foot measured on the diameter and along the axis. For each size thread the length of the thread and the length of engagement of the two mating threads must be held within specified limits. These threads are designed to be self-holding and self-sealing when properly assembled. It is sometimes necessary to turn or to bore a pipe thread with a single-point tool on an engine lathe; however, most pipe threads are cut with a die or with a tap. The thread designation for American Standard Pipe Taper Threads is NPT. A typical thread specification would read ⅜-18-NPT.

There are other pipe threads that are widely used for specialized applications. These include: USA Standard Hose Coupling Threads; National Standard Fire Hose Coupling Screw Thread; American Standard Pipe Threads for Mechanical Joints; American Standard Dryseal Taper Pipe Threads for Pressure-Tight Joints; and British Standard Pipe Threads.

Fig. 12-4. Square Thread.

Ch. 12 SCREW THREADS AND SCREW-THREAD MEASURING

Fig. 12-5. Basic Form of the American Standard Buttress Screw Threads.

Details for the specification of these threads are to be found in the Standards for the specific thread, or in engineering handbooks, such as *Machinery's Handbook*.

American Standard Buttress Screw Thread—The basic form of the thread for the American Standard Buttress Screw Thread is shown in Fig. 12-5. One face, or flank, of this thread is made 7 degrees, with respect to a line perpendicular to the axis, in order to resist heavy loads placed against it. This flank is called the pressure flank of the thread. The opposite flank is 45 degrees, with respect to a line perpendicular to the axis. The basic depth of engagement of the thread is .6 × pitch. A series of nominal major diameters and a series of preferred pitches are provided in the standard. Since Buttress threads are used mainly on specially designed components, specific recommendations concerning combinations of these preferred diameters and pitches are not made.

Buttress threads are used to withstand heavy axial loads such as encountered in the breech mechanism of large guns and airplane propeller hubs. In the past, Buttress threads were not standardized and, as a consequence, variations in small details of the basic form of the thread were common. A metric standard for this thread, that was used in Germany and Italy, did exist. This thread is called *Sägengewinde* in Germany, and *Fillettatura a Dente di Sega* in Italy.

British Standard Whitworth (B.S.W.) and British Standard Fine (B.S.F.) Threads—The basic form of the British Standard Whitworth Thread is shown in Fig. 12-6. This thread was used in Great Britain for general machine construction. The included angle of the thread is 55 degrees and the crest and root are rounded to a radius of .137329 × pitch. The depth is equal to .640327 × pitch. The thread form for the British Standard Whitworth, designated B.S.W., and the British Standard Fine, designated B.S.F., are the same. The B.S.W. series corresponds to the

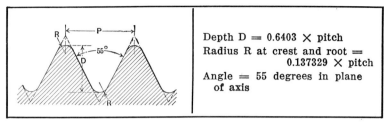

Fig. 12-6. British Standard Whitworth Thread Form.

American coarse series of threads and the B.S.F. corresponds to the fine series. The British Standard specifies a specific number of threads per inch for every nominal diameter.

British Association Standard Thread (B.A.)—This screw-thread system, shown in Fig. 12-7, is recommended for use in preference to the B.S.W. and B.S.F. systems for all screws smaller than ¼ inch. The thread has an included angle of 47.5 degrees. The root and the crest of the thread are rounded. The abbreviation B.A. is used to designate British Association Threads.

British Standard I.S.O. Metric Threads—The design forms of the I.S.O. (International Organization for Standardization) metric internal and external threads are shown in Fig. 12-8. This is the basic metric thread used in the United Kingdom. The principal dimensions are given in the illustration. The depth of the internal thread is .54127 × pitch; for the external thread, the depth is .61344 × pitch. The standard includes a coarse-pitch and a fine-pitch series. The tolerance grades and fundamental deviations for external threads are designated: 8g, 6g, and 4h. For internal threads, the tolerance grades are designated: 7H, 6H, and 5H. A 5H internal thread is used with a 4h external thread to provide a "close" class of fit. The 6H internal thread is used with a 6g external thread to provide a "medium" thread fit. The 7H internal thread fitted to an 8g external thread will result in a "free" fit.

I.S.O. metric screw threads are designated according to the following examples:

$$M5 \times 0.8\text{-}6H$$

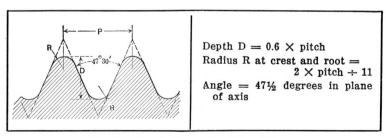

Fig. 12-7. British Association Thread Form.

Ch. 12 SCREW THREADS AND SCREW-THREAD MEASURING 261

Fig. 12-8. Design Forms of Internal and External ISO Metric Thread (Maximum Metal Condition).

This is an internal thread. The M denotes the thread system symbol for I.S.O. metric threads, the 5 denotes the nominal size (diameter) in millimeters, the 0.8 the pitch in millimeters, and the 6H denotes the thread tolerance class symbol.

$$M64 \times 6\text{--}4h$$

This is an external thread. The M denotes the I.S.O. metric thread system, the 64 denotes the nominal size (diameter) in millimeters, and the 4h denotes the thread tolerance class symbol.

In the above two examples the size of the thread is given in terms of the pitch or the distance from a point on a screw thread to a corresponding point on the next thread. This is the custom in the metric system. In the American Standard Unified System the size of the thread is given in terms of threads per inch.

International Metric Thread System—The International Metric Thread System (*Système Internationale*), see Fig. 12-9, was adopted as the standard by most European countries using the metric system, and also by Japan and Russia. (In Great Britain the I.S.O. metric thread is standard.) The thread form is similar to the American Standard except for the depth which is slightly greater. The included angle of the thread is 60 degrees, and the crest has a flat as does the American thread. The width of the flat is one-eighth of the pitch. The root of the thread is rounded which provides a clearance between the root and the crest of the mating thread. The thread depth equals .7035 × pitch, maximum, and .6855 × pitch, minimum. This depth is greater than that of the American thread. The

262 SCREW THREADS AND SCREW-THREAD MEASURING Ch. 12

Max. depth D = 0.7035 × pitch
Min. depth D_1 = 0.6855 × pitch
Root radius
 Max. = 0.0633 × pitch
 Min. = 0.054 × pitch
Angle = 60 degrees in plane of axis.

Fig. 12-9. International Standard Metric Thread.

German metric thread form is like the International Standard; however, the depth is equal to .6495 × pitch.

Measuring and Gaging Screw Threads

Measuring means to determine the exact dimension, while gaging is done to determine if the part is within specified dimensional limits. It is usually more difficult and time consuming to measure an object, including screw threads, than it is to gage the object. Since gaging achieves the desired result of determining the dimensional acceptability of a part, it is usually the preferred method when the cost of the gage per part to be examined is not excessive. Generally, the cost of making gages for low-volume production work is prohibitive; however, standard screw-thread gages are relatively inexpensive and are often kept available in the tool rooms of machine shops and in tool and die shops. Where the volume required is large enough, and where gaging is a requirement of the customer, there can be no question about the availability of thread gages. Measuring methods of determining the acceptability of screw threads are required in making thread gages, when making other threads to exceptionally high standards of accuracy, and when gages are not available. Some of the thread-measuring instruments used now are very elaborate. This chapter will, however, only concern itself with measuring and gaging methods that are commonly used in the machine shop.

Thread Elements to Be Inspected—All of the elements of a screw thread that are listed below are not always inspected. In many cases complete reliance is placed upon an inspection based on GO-NOT GO gages, or even on the "feel" of the fit between the thread and a gage. At other times, threads are measured and gaged with great exactness. The thread elements that are subject to inspection are pitch, pitch diameter, lead, thread form, angle of thread, major diameter, minor diameter, taper, and surface finish.

Thread Plug and Thread Ring Gages—Gaging affords a fast, reliable method of determining the acceptability of external and internal threads. The skill required to use thread gages can, in most cases, be acquired without too much difficulty. One disadvantage of gaging is that part of the thread tolerance must be given to the gage in allowing for a gage-makers tolerance and gage wear. The practical effect of this is that the

Ch. 12 SCREW THREADS AND SCREW-THREAD MEASURING

thread must be made more accurate than called for by the drawing in order to be accepted by the gages. There are often three classes of thread gages used in a shop: *working gages, inspection gages,* and *master gages.* The working gage is to be used on the machine in making the thread. The inspection gage is used by the inspector to accept or reject the finished product. Master gages represent the basic size of the thread. They are made with no allowance for wear and with the minimum gagemakers tolerance. These tolerances and the wear allowances of inspection and working gages are usually so arranged that the inspection gage will accept any thread that the working gage will accept, and will reject any thread that has been rejected by the working gage. In some shops an additional class of gages called *setting gages* is used to set thread ring and snap gages.

Thread ring and snap gages are illustrated in Fig. 12-10. They are usually based on the GO-NOT GO principle. For example, an external thread is accepted if the GO ring gages will screw over the entire length of the thread and if the NOT GO gage cannot be turned onto the thread more than one and one-half turns. The thread is rejected if either the GO gage fails to go on the thread for its entire length, or if the NOT GO gage can be turned for more than one and one-half turns. The GO ring gage should fit onto the thread without binding, or "catching," anywhere along the length of the thread. In some shops some tightness is permitted on the first two threads. *Thread ring gages* for inspecting tapered threads should go on until the back face of the gage is flush with the end of the thread. In inspecting parallel screw-threads an attempt should be made

Courtesy of Colt Industries, Pratt & Whitney Machine Tool Div.
Courtesy of Hanson-Whitney Co.
Courtesy of Greenfield Tap & Die Corp.

Fig. 12-10. Typical thread ring and thread plug-gages.

to "shake" the gage endwise, and to "rock" it, in order to detect any looseness in the fit of the thread gage on the work. Differences in the effort required to turn the gage will indicate the existence of taper or a lead error in the thread. In any case the gage should never be forced over the thread. If it does not go with a light effort it should not be forced to go further. If the gage loosens markedly, after it has been turned along the length of the thread, a more comprehensive check of the back section of the thread is warranted. In this case it is possible for the back section of the thread to be undersize without being detected by the NOT GO gage because it will not pass over the front of a thread which is within the dimensional limits.

A thread ring gage cannot be used to analyze the individual errors that could be present. For example, if the GO gage will not go onto a thread, the difficulty may be due to the pitch diameter being oversize, the minor diameter may be oversize, or there may be dirt in the threads. The GO gage may pass over threads having the major diameter undersize or errors in the thread form. The threads on the NOT GO ring gage have their roots cleared to a greater extent and their crests truncated more than the GO ring gages. This is done in an attempt to assure that any contact between the threads on the gage and the work will be on the sides of the threads and that other conditions will not affect the fit. The NOT GO gage thus is affected only by the pitch diameter of the thread.

Regardless of its limitations, the thread ring gage is a very reliable method of inspecting screw threads. It has one great advantage in that it actually passes over and inspects the entire thread area. The *thread plug gage* also possesses this advantage. GO and NOT GO thread plug gages are used to inspect internal threads much in the same manner as thread ring gages are used to inspect external threads. For an internal thread to be accepted the GO thread plug gage must pass through the entire length of the internal thread and the NOT GO plug gage must not enter the threads for more than one and one-half turns.

The Mating Part as a Gage—A very effective method of gaging a screw thread on certain classes of work is to use a mating part as a gage. In situations where thread gages are not available, and where a thread is being made to fit a particular machine part or a tool, using the work as a gage will bring about the desired fit. In this case extra care should be taken to see that both threads are clean before they are assembled for gaging purposes. Care should also be exercised in examining the "feel" of the thread fit since the lower limit of the thread cannot be verified with a NOT GO gage. The two threads should fit together with no endwise shake and without binding. When no gages are available, and it is inconvenient to use the mating part as a gage, a bolt, cap screw, or a nut, as the case may be—that is known to have accurate threads—can be used as a gage. In many cases this method produces very acceptable results. Like the thread ring or plug gage, the mating part envelops and gages the entire thread area. It is not recommended, however, when threads are to be made to tolerances supplied by the customer.

Ch. 12 SCREW THREADS AND SCREW-THREAD MEASURING 265

Fig. 12-11. Thread snap-gage.

Thread Snap Gages—When the workpiece is turned between centers on a lathe it must be removed in order to try the ring gage. Furthermore a certain amount of time is required to turn the ring gage on and off the work. The thread snap gage, Fig. 12-11, is used somewhat like a micrometer caliper and there is no need to remove the work from the lathe. It has two pairs of anvils that are adjusted by plug type setting gages to GO and NOT GO dimensions. The gage should be carefully aligned with the threads on the workpiece and then passed over the workpiece with a light pressure. It should not be forced because the C frame can spring open enough to pass oversize threads.

Courtesy of Federal Products Corp.
Courtesy of Colt Industries, Pratt & Whitney Machine Tool Div.

Fig. 12-12. Indicating-type thread comparator for inspecting external thread pitch-diameters. Initial setting established with a setting plug.

266 SCREW THREADS AND SCREW-THREAD MEASURING Ch. 12

Another version of the thread snap gage is the *roll thread snap gage*. The gaging surfaces are free-turning rolls that can move laterally in order to align themselves with the threads on the workpiece. The rolls are not actually threads but are annular rings shaped to the form of the thread. They are made to work on the GO-NOT GO principle. Indicating type thread comparators, Fig. 12-12, are similar in some respects to thread snap gages. The thread is placed between the rolls and against a backstop. The dial test indicator will then read the variation in the pitch diameter, directly. The comparator is set to master plug gages. Care must be taken in order to make certain that the backstop is set so that the thread-gaging rolls contact on the true diameter of the workpiece. On some indicating types of thread comparators, three sets of rolls, instead of two, are used. Two lower sets of rolls are used to nest the thread of the workpiece in the manner of a V-block. An advantage of the indicating type of thread gage is its ability to check out-of-round, and taper, in the thread. A roll-type thread gage for inspecting tapered threads is shown in Fig. 12-13.

Optical Comparators—The optical comparator, Fig. 12-14, can be used to make precise measurements of screw threads. It is possible to measure the thread form, thread angle, major diameter, minor diameter, pitch diameter, pitch, and lead error in one step. The condition of the surface of the thread can be observed by using the surface illuminator.

The optical comparator casts a beam of high-intensity light against the thread. The shadow cast by the thread profile is magnified and projected onto a ground-glass screen where it can be focused to form an enlarged silhouette of the thread. The range of magnifications available is usually from 5× to 250×. The thread to be measured is held in position by a

Courtesy of Colt Industries, Pratt & Whitney Machine Tool Div.

Fig. 12-13. (Left) Roll-type thread gage for checking tapered thread. (Right) Roll-type gage for checking major diameter of a tapered thread.

Ch. 12 SCREW THREADS AND SCREW-THREAD MEASURING 267

Courtesy of Jones & Lamson Division of Waterbury-Farrel Company

Fig. 12-14. Optical Comparator.

V-block, or between centers, as in B, Fig. 12-15. The table of the projector can be moved in three directions: perpendicular to the lens, parallel to the lens, and vertically. The table movements can be very accurately

Courtesy of Jones & Lamson Division of Waterbury-Farrel Company

Fig. 12-15. A (Left). Projection of a 13-pitch thread at 50 magnifications. B (Right). Measuring the lead of a screw thread on an optical comparator.

determined by a micrometer and end-measuring rods. The hairlines on the viewing screen can be rotated 360 degrees to measure angles. The angle can be measured to an accuracy of one minute of arc by a vernier scale. Standard thread comparison charts can be obtained as shown at A, Fig. 12-15. The chart may be made of plate glass, substituted for the viewing screen supplied with the projector; or, it may be made from a vinyl plastic which is placed over the viewing screen and held in place by chart clips which are attached to the projector.

Toolmakers Microscope—Toolmakers microscopes can be used to measure the thread form, thread angle, major diameter, minor diameter, pitch diameter, pitch, and lead error. The table can be moved in two perpendicular directions by precision lead-screws. The lead screws are actuated by micrometer thimbles that have graduations reading to an accuracy of .0001 inch. The eyepiece can be equipped with standard reticles that have hairlines in the form of the thread.

Three-Wire Method of Thread Measurement—The three-wire method of measuring the pitch diameter of a thread utilizes three precision wires, or pins of exactly the same diameter, which are positioned in the threads as shown in Fig. 12-16. The size of the pitch diameter is determined by measuring over the wires with a micrometer caliper and by applying this measurement in an appropriate formula. The three-wire method is the accepted method of measuring the pitch diameter on precision threads, such as thread gages, instrument threads, etc. It is not customarily employed in ordinary commercial work because thread gages are faster and easier to use. The three-wire method can be used to compare the size of two threads without the use of formulas. For example, suppose that a certain external thread must be reproduced and that there is no gage or

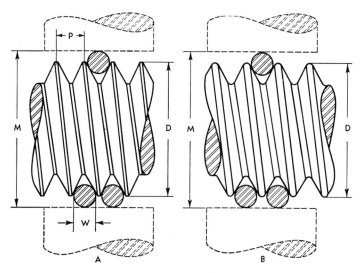

Fig. 12-16. Measuring pitch-diameter by the three-wire method.

Ch. 12 SCREW THREADS AND SCREW-THREAD MEASURING

mating thread available with which to inspect the new thread; one method would be to make a gage to fit the original thread; however, this would be time consuming and expensive. Another method would be to measure the original part with three wires in several places to obtain an accurate representative measurement, M. The new thread is then made to the same size as M when measured over the three wires that were used to measure the original thread. Although this cannot be classified as the three-wire method, ordinary calipers can be modified to compare the sizes of screw threads in a similar manner. The outside calipers are modified by filing the points of the legs round, so that they contact only the sides of the thread and not the roots.

The following notation applies to the formulas for measuring threads by the three-wire method:

D = Basic major outside diameter
N = Number of threads per inch = $1 \div P$
E = Pitch diameter
M = Dimension over wires
P = Pitch = $1 \div N$
W = Wire or pin diameter
A = One-half of the included thread angle in the axial plane
A_n = One-half of the included thread angle in the normal plane, or the plane perpendicular to the sides of the thread.
B = Lead angle at the pitch diameter
L = Lead of thread
T = Width of thread in the axial plane at diameter E.

Formulas 12-1, 12-2, and 12-3, given here, are simplified formulas that do not include the effect of the lead angle. They are, however, sufficiently accurate for many applications involving single-thread screw threads. These equations are used for threads having the following thread forms: American Standard Unified, American National, British I.S.O. Metric, International Metric.

To find M, when the standard major diameter is known:

$$M = D - (1.5155 \times P) + (3 \times W) \qquad (12\text{-}1)$$

To find M, when the pitch diameter E, is known:

$$M = E - (.86603 \times P) + (3 \times W) \qquad (12\text{-}2)$$

To find the best wire size:

$$W = .57735 \times P; \text{ or } W = .57735 \div N \qquad (12\text{-}3)$$

The effect of the lead angle of the thread is included in Formula 12-4. It can be used to calculate the pitch diameter of any thread form having symmetrical sides (American Standard Unified and National, B.A., B.S.W., British I.S.O. Metric, International Metric). It is recommended

for calculating the pitch diameter for Acme and for worm threads that do not have an excessively large lead angle which is the case for most single-thread screws. Formula 12-4 is a simplification of one that is much more cumbersome and difficult to apply; however, the wire diameter used in Formula 12-4 must be the exact wire diameter obtained in Formula 12-4a. This, in effect, means that special wires of the size calculated from Formula 12-4a must be used.

$$M = E + W\,(1 + \sin A_n) \qquad (12\text{-}4)$$

$$W = \frac{T \times \cos B}{\cos A_n} \qquad (12\text{-}4a)$$

$$\tan A_n = \tan A \times \cos B \qquad (12\text{-}4b)$$

$$\tan B = L \div 3.1416 \times E \qquad (12\text{-}4c)$$

$$T = .5 \times P \qquad (12\text{-}4d)$$

Formula 12-4 combines simplicity with a degree of accuracy that meets all but the most exacting requirements, particularly for lead angles below 8 or 10 degrees and for the higher thread angles. Additional equations for B.S.W. and B.S.A. threads are given below. These equations do not compensate for the lead angle; however, they do not require the use of special wires. Formula 12-7 gives the recommended wire size, although deviations from this size can be made.

Formulas for the British Standard Whitworth Thread Form (B.S.W.) are:

$$M = D - (1.6008 \times P) + (3.1657 \times W) \qquad (12\text{-}5)$$

$$M = E - (.9605 \times P) + (3.1657 \times W) \qquad (12\text{-}6)$$

$$W = .56368P; \text{ or } W = .56368 \div N \qquad (12\text{-}7)$$

Equations for the British Association Thread Form (B.A.) are:

$$M = D - (1.7363 \times P) + (3.4829 \times W) \qquad (12\text{-}8)$$

$$M = E - (1.1363 \times P) + (3.4829 \times W) \qquad (12\text{-}9)$$

$$W = .5 \times P \times \sec 23°45' = .5462P \qquad (12\text{-}10)$$

The application of the formulas for the three-wire method can best be shown by the following examples.

Example 12-1:

A $1\frac{1}{2}$-12 UNF-3C thread is to be measured by the three-wire method. Calculate the dimension, M, over the three wires.

$$\begin{aligned} W &= .57735 \div N \\ &= .57735 \div 12 \\ &= .04811 \text{ inch} \end{aligned} \qquad (12\text{-}3)$$

Ch. 12 SCREW THREADS AND SCREW-THREAD MEASURING

The nearest available wire size is .050 inch.

$$M = D - (1.5155 \times P) + (3 \times W) \quad (12\text{-}1)$$
$$= 1\tfrac{1}{2} - (1.5155 \times \tfrac{1}{12}) + (3 \times .050)$$
$$= 1.5237 \text{ inches.}$$

The thread should be cut until the measurement over the three wires is 1.5237 inches.

Example 12-2:

According to the thread standards (*Machinery's Handbook*, 18th Ed., p. 1226) for Unified Threads, the pitch diameter for the 1½-12 UNF-3A thread must be 1.4459 inches maximum and 1.4411 inches minimum. Calculate the maximum and minimum measurement M, over the three wires.

$$M = E - (.86603 \times P) + (3 \times W)$$
$$= E - (.86603 \times \tfrac{1}{12}) + (3 \times .050)$$
$$= E - .07217 + .150 = E + .0778 \quad (12\text{-}2)$$

Thus:

$$M_{\max} = 1.4459 + .0778 = 1.5237 \text{ inches}$$
$$M_{\min} = 1.4411 + .0778 = 1.5189 \text{ inches}$$

Example 12-3:

The pitch diameter, E, of a ½-10 Acme-2G thread is to be .436 inch and the lead, L, is .100 inch. Calculate the diameter of the wire, W, that must be used to measure this thread by the three-wire method and calculate the measurement, M, over the three wires.

$\tan B = L \div 3.1416 \times E = .100 \div 3.1416 \times .436 = .07301$
$B = 4°11'$ $\qquad\qquad (12\text{-}4c)$

$\tan A_n = \tan A \times \cos B = \tan 14°30' \times \cos 4°11'$
$= .25862 \times .99733 = .25793$ $\qquad (12\text{-}4b)$

$A_n = 14°28'$

$T = .5 \times P = .5 \times .100 = .050$ inch $\qquad (12\text{-}4d)$

$$W = \frac{T \times \cos B}{\cos A_n} = \frac{.050 \times \cos 4°11'}{\cos 14°28'}$$
$$= \frac{.050 \times .99733}{.96829} \qquad (12\text{-}4a)$$

$W = .0515$ inch

$M = E + W(1 + \sin A_n) = .436 + .0515(1 + \sin 14°28') \quad (12\text{-}4)$
$= .436 + .0515(1 + .24982)$

$M = .5004$ inch

The accuracy of the three-wire method of thread measuring is affected by the accuracy of the measuring wires, or pins. If the accuracy of the pitch diameter of the screw thread gage to be measured is within .0001 inch, it is necessary to know the diameter of the wires within .00002 inch. Each wire should be round, within .00002 inch, and should be straight within the same amount over any quarter-inch section. A set of three wires should have the same diameter within .00002 inch; moreover, this common diameter should be within the "best size" for any given pitch. The "best size" is one that makes contact at the pitch line at one-half of the thread depth. This is where any given error in the thread angle will have the least effect upon the measurement over the wires. If the proper precautions are taken, regarding wire accuracy and contact pressure of the micrometer, it should be possible to check plug gages within an accuracy of .0001 inch. If the wire diameters are accurate to only .0001 inch, then the pitch-diameter measurement is not likely to be more accurate than .0003 inch. This, however, may be accurate enough for many classes of work.

Other sources of inaccuracy, when measuring threads by the three-wire method, are errors in thread form and in lead. The three-wire method will not detect errors in thread form; and the errors in the thread form will cause the calculated values of the measurement over the wires, M, or of the pitch diameter, E, to be incorrect. Errors in lead do not cause errors in the measurements or in the calculations. The error in lead cannot be detected by the three-wire method of measuring screw threads, thus other methods must be used to check the lead.

Screw Thread Micrometer—A screw-thread micrometer may be used for measuring the pitch diameter. This micrometer has a fixed anvil which is V-shaped, to fit over the thread, while the spindle is cone shaped at the end in order that it can enter the space between two threads. (See Fig. 12-17.) The anvil and the spindle make contact with the sides of the thread, enabling the pitch diameter to be determined. The cone-shaped point of the measuring screw is slightly rounded so that it will not bear at the bottom of the thread. There is also sufficient clearance at

Fig. 12-17. Screw-thread micrometer.

Ch. 12 SCREW THREADS AND SCREW-THREAD MEASURING

the bottom of the groove in the V-shaped anvil to prevent it from touching at the top of the thread. When the micrometer is closed and the 60 degree point B, and anvil A, are together (as shown by the diagram), the zero on the micrometer barrel represents a line drawn through the plane X-Y, and, if the micrometer is opened, say, .5 inch, this reading represents the distance that two such planes are apart. Hence, the micrometer reading represents the pitch diameter. The thread micrometer does not give an absolutely accurate pitch-diameter measurement as there is a slight error which depends upon the lead angle of the thread. Usually, this error is not large enough to be of practical importance. There is no error, however, when the thread micrometer is set to a standard gage of the plug type and is then used to compare the measurement over the plug with the measurement of the thread on the workpiece. Thread micrometers are made to cover a very limited range of pitches, or threads per inch, which is governed by the pitch of the V-shape grooves in the anvil.

QUESTIONS

12-1. What are the major, minor, and pitch diameters of a screw thread?

12-2. What is the importance of the pitch diameter?

12-3. What is the difference between the pitch and the lead of a screw thread?

12-4. a. Calculate the pitch of a thread having 18 threads per inch.
b. Calculate the number of threads per inch for a thread having a pitch of .125 inch.

12-5. What is the maximum material limit of a screw thread?

12-6. What is the minimum material limit of a screw thread?

12-7. What is the lead angle of a screw thread?

12-8. Make a sketch of an American Standard Unified Thread Form.

12-9. What are the differences between the American Standard Unified Thread Form and the American National Thread Form?

12-10. What is the American Standard Unified Thread System?

12-11. Name all of the different thread series that are part of the American Standard Unified Thread Series.

12-12. What is a thread class?

12-13. Give the nominal size, threads per inch, pitch, series, and class for the following threads:

$\tfrac{5}{16}$-18 UNC-2A \qquad $\tfrac{5}{8}$-20 UN-3A
$\tfrac{3}{8}$-24 UNF-3B \qquad $\tfrac{7}{8}$-16 UN-1A
$\tfrac{1}{2}$-14 UNS-2B \qquad 1-8 UNC-2B
$\tfrac{5}{8}$-24 UNEF-2A \qquad 4-6 UN-2A
6-32 UNC-2A \qquad $\tfrac{5}{8}$-11 NC-2

12-14. What are interference fit threads and when are they used?

274 SCREW THREADS AND SCREW-THREAD MEASURING Ch. 12

12-15. What are the differences among the three types of American Standard Acme Screw Threads?

12-16. Give the nominal size, threads per inch, pitch, and the class of the following Acme threads:

 1-5 Acme-2G
 1½-.25p-0.50L Acme-3G
 3-2-Stub Acme-2G

12-17. Name five types of pipe threads.

12-18. Why are the American Standard Taper Pipe Threads tapered?

12-19. What is a Buttress thread and why is it used?

12-20. Name four thread systems commonly used in Great Britain.

12-21. What thread system is used in Europe?

12-22. What is the difference between measuring a screw thread and gaging it?

12-23. Name the elements of a screw thread that are inspected.

12-24. How should a GO-NOT GO set of ring gages be used to inspect a screw thread?

12-25. What are the advantages and the disadvantages of gaging, as compared to measuring, screw threads?

12-26. How is the mating part used as a gage?

12-27. What is the advantage of a thread snap gage? What is its disadvantage?

12-28. What is the advantage of using an optical comparator to measure a screw thread?

12-29. How are threads measured by the three-wire method?

12-30. What are the limitations of the three-wire method?

12-31. Calculate the measurement, M, over three wires for measuring a ¾-10 UNC-3A thread. Assume that the best wire size is available.

12-32. Calculate the measurement, M, over three wires for measuring a ¼-20 UNC-2A thread. Assume that the best wire size is available.

12-33. Calculate the wire size, W, and the measurement, M, for a 2½-3 Acme-2G thread which has a lead, L, of .3333 inch and a pitch diameter equal to 2.3105 inch.

CHAPTER **13**

Cutting Screw Threads on a Lathe

Engine lathes are designed with features that are provided primarily for cutting screw threads. The gear ratios in the feed-thread gear box are selected to enable the lathe to cut a sequence of precise screw-thread leads. Most engine lathes are made so that these gear ratios can be modified by change gears to enable metric threads, or threads having special leads, to be cut. The lead screw, the split nut (or half nut), and the thread chasing dial are built onto the lathe for the exclusive purpose of cutting screw threads. The compound rest has many applications among which is to feed the cutting tool into the work at the proper angle when cutting screw threads. The degree of precision that is built into the lathe enables it to cut very accurate threads.

Thread cutting on a lathe is often called thread chasing. Practically every kind of thread can be cut on an engine lathe. A single-point cutting tool is used to cut screw threads on engine lathes. This tool has the advantages of low initial cost, is easily ground and resharpened, and is simple to set up on the lathe. The low tooling and set-up costs provide an advantage for producing screw threads on an engine lathe when the number of parts to be made is small. Even with larger production volumes, the lathe sometimes has an inherent advantage. For example, threads are cut on parts that are turned on a lathe in order to avoid an additional setup on another machine. Other methods of cutting screw threads are: Chasing threads with a threading die, tapping, thread milling, thread rolling, and thread grinding. Although each of these methods has its special advantages on certain classes of work, there is still much thread cutting done on the lathe.

The Single-Point Thread-Cutting Tool

The start of a good thread is the thread-cutting tool. It must have the correct shape, or profile, in order to cut an accurate thread. Usually this shape corresponds exactly with the shape of the thread to be cut. Accurate thread profiles can be ground on tool and cutter-grinding machines. While it is possible to grind a thread-cutting tool by hand on a pedestal grinder, the tools for extremely accurate threads should always be ground on the cutter and tool-grinding machines. Cemented-carbide thread-cutting tools are usually purchased already formed to the required shape, as shown in Fig. 7-7. High-speed-steel thread-cutting tools are usually

ground to the required shape. Whether the thread-cutting tool is ground to the required shape on a cutter and tool-grinding machine, or by hand, on a pedestal grinder, it should be carefully inspected before it is used.

The design of a high-speed-steel cutting tool for cutting American Standard Unified and National Thread Forms is shown in Fig. 13-1. This tool has a 60-degree included angle and the width of the flat, F, is made appropriate to the pitch of the thread to be cut. View A, is a top view of the tool showing the shape, or profile. An alternate shape is also shown. The advantage of the alternate shape is that it reduces the surface area on the flanks of the tool which must be ground to obtain this shape, thereby reducing the area in contact with the grinding wheel. This, in turn, reduces the heat generated by the grinding wheel. It also makes it easier to control the tool while grinding, in order to obtain the desired profile. View B, is a side view of this tool, showing the angle made by the flat. This angle should be 10-to-17 degrees depending upon the hardness of the metal being cut. The right side of the tool is shown at C. This view is a true auxiliary projection which makes the tool appear to be "upside down." It is an end view of the tool sighting, parallel to the left side cutting edge and it shows the relief angle below this edge. The magnitude of the left side-cutting edge relief angle should be 10 to 17 degrees when cutting right-hand screw threads. The lower value should be used when cutting hard materials, and for cutting soft materials the size of this angle is increased to 17 degrees. View D is a true auxiliary view obtained when sighting parallel to the right side-cutting edge of the threading tool. It shows the right side-cutting edge relief angle which should be 8 to 12 degrees when cutting right-hand screw threads. The face of the thread-cutting tool is ground flat, or without a rake angle. It should, however, be ground smooth in order for all the cutting edges to be sharp. When cut-

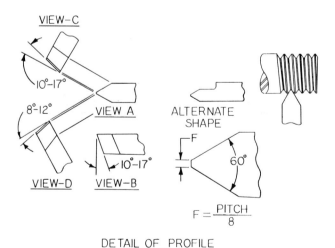

Fig. 13-1. Thread-cutting tool for American Standard Unified and National Threads.

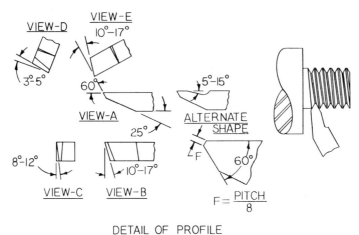

Fig. 13-2. Offset thread-cutting tool for American Standard Unified and National Threads.

ting large threads (say, 5 threads per inch, and larger), the threads are sometimes rough cut with a cutting tool having a side-rake of 8 to 12 degrees, and having an included angle of 50 degrees, instead of 60 degrees. The large thread is, however, finished to size with a tool having the correct included angle and no rake angle.

An offset thread-cutting tool is shown in Fig. 13-2. This threading tool is recommended when it is necessary to cut close to a shoulder or close to a lathe dog. The offset thread-cutting tool is set at an angle, as shown. This allows it to cut ahead of the cross slide and compound rest to prevent these parts from bumping into large shoulders or the lathe dog. View A is the top view of the offset thread-cutting tool, showing its profile. The alternate shape shows a 5- to 15-degree "hook," or relief, ground into the right side-cutting edge. The purpose of this relief is to reduce the area of the flank under the right side-cutting edge in order to facilitate the grinding of this surface. View B, is of the left side of this tool, showing the relief angle under the flat at the nose of the tool. It should be mentioned, however, that View B does not show this angle in its true projection. An end view showing the right side-cutting edge relief angle in its true projection is shown at E. The auxiliary view, D, is a view from the end of the tool, looking along the edge which is 25 degrees, with respect to the shank. A clearance angle of 3 to 5 degrees is customarily ground below this edge. Note that both views D and E appear to be "upside down," which is in accord with drawing practice.

The relief angles below the cutting edge that are recommended for cemented-carbide tools should be less than those recommended for high-speed steel. The reason for this is that cemented carbides, while being harder than high-speed steel, are also more brittle, and the cutting edge requires the additional support provided by having smaller relief angles.

Thread-cutting tools must be ground to have the form of the thread to be cut. The relief angles below the cutting edges, however, should be essentially the same as recommended in Figs. 13-1 and 13-2. The profiles of thread-cutting tools and other thread forms are shown in Figs. 13-3 through 13-6.

The form of the thread-cutting tool must be inspected for accuracy. Typical gages for checking the thread form are shown in Fig. 13-7. A center gage, such as is used for checking lathe centers, is shown at A. This is used for checking the 60-degree included angle of an American Standard Unified and National Thread Form. Form gages such as those in Fig. 13-7 should be held on the same plane as the face of the thread-cutting tool, shown at A. If the tool is made to fit the gage when the gage is held as shown by the dotted line, the included angle will be incorrect. The center gage has no provisions for checking the width of the flat at the nose of the tool. This can be done with the circular thread-gage shown at B. Opposite each notch around the circumference of the gage is stamped the number of threads per inch for which the notches have the correct form. The tool is first ground to the 60-degree angle. This angle is checked by

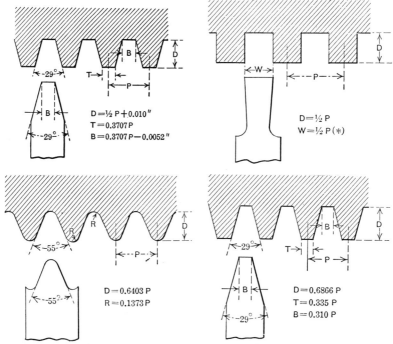

Fig. 13-3. (Upper left) American Standard Acme thread-cutting-tool profile. Fig. 13-4. (Upper right) Square thread-cutting-tool profile. Fig. 13-5. (Lower left) British Standard Whitworth thread-cutting-tool profile. Fig. 13-6. (Lower right) 29-degree worm thread-cutting-tool profile.

Fig. 13-7. Gages for inspecting thread-cutting tools.

the notch at d which has a sharp V to permit the tool to be ground to a sharp point. The correct width is then ground as gaged by the notch corresponding to the threads per inch to be cut by the tool.

An Acme gage is shown at C, Fig. 13-7. This also has notches for different pitches or threads per inch. The thread-cutting tool is first ground to the correct angle using the 29-degree notch at the end of the gage, as shown at g. The width at the end of the cutting tool is then ground—using the appropriate shallow notch to gage this dimension—as at h. The numbers opposite each notch represent the number of threads per inch. The angle between the side of the tool and end-width may be tested as illustrated at j. This gage may also be set square with the work by placing one edge of the gage against the turned surface and adjusting the tool until it coincides with the gage, as indicated by the dotted lines at j.

The end width of the Acme tool may also be measured with a gear-tooth vernier, as shown in Fig. 13-8. The distance, A, is usually made to equal .250 inch by setting the vertical vernier scale to this dimension. The width of the flat is then equal to the caliper-jaw, vernier scale reading minus $2 \times .250 \times$ Tangent $14°30'$, or $2 \times .250 \times .2586 = .1293$ inch. This can be simplified by using the following rule: The width of the end of the Acme thread tool is equal to the reading of the caliper-jaw, vernier scale minus .1293 inch, when the vertical scale is set to .250 inch and the caliper jaws touch on the side of the tool. It is, of course, assumed that the included angle of the tool is 29 degrees. For example, assume that the distance A, Fig. 13-8, is .250 and when the caliper jaws are touching the sides of the tool, the vernier reading is .315 inch. The width of the end of the Acme thread-cutting tool is then $.315 - .1293 = .1857$ inch.

The optical projector, Fig. 12-14, can also be used to measure thread-cutting tools. This is a very accurate method and its use is recommended.

Fig. 13-8. Measuring the end width of an Acme thread-cutting tool with vernier gear-tooth caliper.

Likewise, a toolmakers microscope can be used for this purpose. The instruments provide the best method of checking thread-cutting tools when they are available.

Machine Setting—Change Gear Ratios

Almost all modern engine lathes are equipped with a feed-thread gear box (Fig. 6-7). The lathe is set to cut the required number of threads per inch by simply placing a few levers or handles in their correct position according to the instructions given on the chart attached to the feed-thread gear box.

Many older lathes were not equipped with a feed-thread gear box; instead, they had change gears such as shown in Fig. 13-9. The calculation of the change gears required to cut a given thread introduces the principles of gear ratios and shows, in principle, how the feed-thread gear box works. For this reason a study of the calculation of change gears is worthwhile.

First, one must understand why a lathe cuts a certain number of threads per inch and how this number is changed by the use of different gears. As the carriage and the cutting tool are moved by the lead screw, which is connected by gearing with the lathe spindle, the number of threads per inch that are cut depends, in every case, upon the number of turns the work makes while the lead screw is moving the carriage one inch. If the lead screw has six threads per inch, it will make six revolutions while the carriage and the thread tool travel one inch along the piece to be threaded. If change gears a and c (A, Fig. 13-9) are so proportioned that the spindle makes the same number of revolutions as the lead screw in a given time, it is evident that the tool will cut six threads per inch because the spindle has made six revolutions while the carriage and the tool have traveled one inch. If the spindle revolved twice as fast as the lead screw it would make twelve turns while the tool moved one inch, and conse-

quently, twelve threads per inch would be cut. To obtain this difference in speed it is necessary to use a combination of gearing that will cause the lead screw to revolve only once while the lathe spindle and work make *two* revolutions. On lathes equipped with a feed-thread gear box this is done by the gear ratios inside the box. On lathes not so equipped, this is done by selecting appropriate change gears.

For calculating change gears, the lead number of the lathe must first be obtained. The lead number of the lathe, also called the lathe screw constant, is defined as the number of threads per inch cut by the lathe when gears of the same size are placed on the lead screw and the spindle stud. (See Fig. 13-9.) This can be determined experimentally by placing two, equal-size gears on the lead screw and stud, or this number is given on a chart accompanying the gears. The gears for any thread are then calculated by the following equation:

$$\frac{\text{Lead number of lathe}}{\text{Threads per inch to be cut}} = \frac{\text{Number of teeth on spindle-stud gear or driving gear}}{\text{Number of teeth on lead-screw gear or driven gear}} \quad (13\text{-}1)$$

Fig. 13-9. A. Lathe with simple gearing for thread cutting. B. Lathe with compound gearing for thread cutting.

Example 13-1:

A ¾-10 UNC-2A thread is to be cut on a lathe having a lead number equal to 6. Calculate the change gears required:

$$\frac{\text{Lead number of lathe}}{\text{Threads per inch to be cut}} = \frac{6}{10} = \frac{6 \times 4}{10 \times 4} = \frac{24}{40} = \frac{\text{Driving gear}}{\text{Driven gear}}$$

Note that the ratio $\frac{6}{10}$ has not changed because it was multiplied by $\frac{4}{4}$ which is equal to 1. Furthermore, the number of teeth on the intermediate, or idler gear, *b*, which connects the stud and lead-screw gears, is not considered because it does not affect the ratio between gears *a* and *c*; however, it does affect their direction of rotation. The gearing as shown is for cutting right-hand screw threads. One additional idler gear (not a compound gear as at B, Fig. 13-9) is used between gears *a* and *c* when the direction of rotation of the lead screw must be reversed in order to cut left-hand threads. The second idler gear, like the first, affects only the direction of rotation and not the ratio of gears *a* and *c*.

Sometimes the change gears cannot be arranged, as shown at A, Fig. 13-9, because the required gears to cut the desired threads per inch are unavailable. In this case it is necessary to use four gears, or compound gearing, as shown at B. The calculations required to determine the size of the compound gears are shown by the following example:

Example 13-2:

A ⁵⁄₁₆-24 UNF-2A thread is to be cut on a lathe that has a lead number equal to 6. Calculate the size of the change gears required to cut this thread by using a compound-gear ratio.

$$\frac{\text{Lead number of lathe}}{\text{Threads per inch to be cut}} = \frac{6}{24} = \frac{2 \times 3}{4 \times 6} = \frac{(2 \times 20) \times (3 \times 10)}{(4 \times 20) \times (6 \times 10)}$$

$$= \frac{40 \times 30}{80 \times 60} = \frac{\text{Driving gear}}{\text{Driven gear}}$$

Note again, that the original ratio $\frac{6}{24}$ is never changed. In order to calculate compound gears the ratio was changed from $\frac{6}{24}$ to $\frac{2\times 3}{4\times 6}$ and each term is then multiplied by a ratio equal to 1; i.e. $\left(\frac{20}{20} = 1 \text{ and } \frac{10}{10} = 1\right)$. The position of the gears should be such that the driving gears act as drivers and the driven gears follow the drivers. At B, Fig. 13-9, gear *a* is a driving gear that drives gear *b*, and gear *c* is a driving gear that drives gear *d*. The gears may be positioned in any of the following combinations in order to cut 24 threads per inch:

CUTTING SCREW THREADS ON A LATHE

	Driving Gear a	Driven gear b	Driving Gear c	Driven Gear d
	Teeth per gear			
Combination 1	40	80	30	60
Combination 2	40	60	30	80
Combination 3	30	80	40	60
Combination 4	30	60	40	80

Observe that the 40-and 30-tooth gears are always driving gears and that the 80-and 60-tooth gears are always driven gears.

Metric threads can be cut on lathes equipped with a lead screw having English dimensions by using a pair of suitable translating gears in the compound-gear train. Metric threads are usually specified by the pitch, in terms of millimeters (mm). This pitch must be converted into threads per inch. There are 2.54 centimeters (cm) per inch and there are 10 mm per cm.

$$\text{Threads per cm} = \text{threads per mm} \times 10 = \frac{1}{\text{pitch in mm}} \times 10$$

$$\text{Threads per cm} = \frac{10}{\text{pitch in mm}} \qquad (13\text{-}2)$$

$$\text{Threads per inch} = \text{Threads per cm} \times 2.54$$

$$= \left(\frac{10}{\text{pitch in mm}}\right) \times 2.54$$

Applying this relationship to Equation 13-1:

$$\frac{\text{Lead number of lathe}}{\text{Threads per inch to be cut}} = \frac{\text{Driving gear}}{\text{Driven gear}}$$

$$\frac{\text{Lead number of lathe}}{\left(\dfrac{10}{\text{pitch in mm}}\right) \times 2.54} = \frac{\text{Lead number of lathe} \times 50}{\left(\dfrac{10}{\text{pitch in mm}}\right) \times 2.54 \times 50}$$

$$= \frac{\text{Driving gear}}{\text{Driven gear}}$$

$$\frac{\text{Lead number of lathe} \times 50}{\left(\dfrac{10}{\text{pitch in mm}}\right) \times 127} = \frac{\text{Driving gear}}{\text{Driven gear}} \qquad (13\text{-}3)$$

The 50-tooth and the 127-tooth gears are called the translating gears. The following example shows how Formula 13-3 is used:

Example 13-3:

Determine the change gears required to cut a metric thread having a pitch of 2 mm on a lathe having a lead number equal to 6.

$$\frac{\text{Lead number of lathe} \times 50}{\left(\dfrac{10}{\text{pitch in mm}}\right) \times 127} = \frac{6 \times 50}{\left(\dfrac{10}{2}\right) \times 127} = \frac{6 \times 50}{5 \times 127}$$

$$= \frac{(6 \times 6) \times 50}{(5 \times 6) \times 127} = \frac{36 \times 50}{30 \times 127} = \frac{\text{Driving gears}}{\text{Driven gears}}$$

Metric threads can also be cut on most lathes equipped with a quick-change gearbox by replacing the quadrant gear (refer to Fig. 6-4) with the two translating gears to form a compound-gear train. The 127-tooth gear is engaged with the headstock stud-gear and the 50-tooth gear drives the gear attached to the quick-change or feed-thread gear box. With the translating gears in place the lathe is set to cut metric threads by simply setting the levers and handles on the quick-change gear box, to the threads per inch corresponding to the threads per cm, as calculated by Formula 13-2. It should be emphasized that the lathe must be constructed so that the translating gears can be made to fit in place of the quadrant gear, otherwise metric threads cannot be cut. Most lathes are designed to permit this to be done.

The gearing for cutting worm-threads is done in a similar way. Worm threads are designed to work in engagement with worm gears. The teeth of worm gears are specified in terms of diametral pitch which is equal to the number of teeth on the gear divided by the pitch diameter. The pitch of the worm thread is equal to the distance between each worm-gear tooth which is called the circular pitch of the gear. The circular pitch is equal to π (pi) divided by the diametral pitch, where π is equal to 3.1416 or very nearly $\dfrac{22}{7}$. Thus, for worm threads:

$$\text{Circular pitch of worm gear} = \text{Pitch of worm thread}$$

$$= \frac{22}{7 \times \text{Diametral pitch}}$$

$$\text{No. threads per inch of worm} = \frac{1}{\text{pitch}} = \frac{1}{\dfrac{22}{7 \times \text{Diametral pitch}}}$$

$$= \frac{7 \times \text{Diametral pitch}}{22}$$

This ratio can now be incorporated into the change-gear formula.

$$\frac{\text{Lead number of lathe}}{\text{Threads per inch to be cut}} = \frac{\text{Lead number of lathe}}{\dfrac{7 \times \text{Diametral pitch}}{22}}$$

$$\frac{\text{Lead number of lathe} \times 22}{\text{Diametral pitch} \times 7} = \frac{\text{Driving gears}}{\text{Driven gears}}$$

Example 13-4:

Calculate the change gearing required to cut the worm thread for a 5-diametral-pitch worm gear (single thread) if the lead number of the lathe is 4.

$$\frac{\text{Lead number of lathe} \times 22}{\text{Diametral pitch} \times 7} = \frac{4 \times 22}{5 \times 7} = \frac{88}{35} = \frac{\text{Driving gear}}{\text{Driven gear}}$$

If these gears are not available, compound gears can be used.

$$\frac{\text{Lead number of lathe} \times 22}{\text{Diametral pitch} \times 7} = \frac{4 \times 22}{5 \times 7} = \frac{(4 \times 6) \times (22 \times 3)}{(5 \times 6) \times (7 \times 3)}$$

$$= \frac{24 \times 66}{30 \times 21} = \frac{\text{Driving gears}}{\text{Driven gears}}$$

Worm threads could be cut on lathes equipped with a quick-change gearbox by replacing the quadrant gear (Fig. 6-4) with a 66-tooth driving gear and a 21-tooth driven gear provided that these gears can be made to engage the stud and the gearbox gear. Where this can be done, the quick-change gearbox is set to cut the number of threads corresponding to the diametral pitch and the translating gears (66-tooth and 21-tooth) are inserted in place of the quadrant gear in order to cut the worm thread.

Machine Setting—Compound Rest

In setting up the machine for cutting screw threads, the compound rest should be positioned to feed the cutting tool into the work at an angle, so that only one cutting edge forms a chip. For American Standard Unified and National threads, or any thread having a 60-degree included angle, this angle should be 60 degrees with respect to the axis of the lathe or 30 degrees with respect to the cross slide. Some machinists prefer to set the compound rest at 29 degrees, with respect to the cross slide. For Acme threads the compound rest should be set at $14\frac{1}{2}$ degrees with respect to the cross slide, or $75\frac{1}{2}$ degrees with the axis of the lathe.

Cutting Speed for Thread Cutting

Theoretically, the same cutting speed can be used to cut screw threads on an engine lathe as is recommended for turning. In practice, threads can sometimes be cut at this speed. There are limitations, however, that are imposed by the design of the lathe, the workpiece, the lathe operator, and sometimes by the setup of the workpiece on the lathe, that must be weighed in arriving at a lathe spindle speed. Some lathes are easier to manipulate for thread cutting than others. For example, it is easier to engage the split nut over the lead screw on a 16-inch lathe with a coarse pitch thread than on a smaller lathe with a finer-pitch lead screw thread. The cross slide can also be moved more rapidly on the 16-inch lathe because of the coarser-pitch thread on the cross-feed screw. Fine-pitch threads on the workpiece can usually be cut at a faster speed than coarse-pitch threads. The spindle speed must often be reduced when cutting deep

threads such as Acme, or square, threads. Often, the speed must be reduced when the thread ends close to the shoulder. In some cases the setup of the workpiece necessitates a reduction in speed. An example of this can be seen in Fig. 13-18. The skill of the operator and his familiarity with a given machine must also be taken into account.

A learner should start out by using a slow spindle speed for cutting threads. As skill and confidence are gained, the spindle speed of the lathe can be increased. It must be emphasized that thread cutting is not necessarily a slow-speed operation. The spindle speed should be as fast as conditions will permit, up to the limit imposed by the cutting speed recommended for the material being cut. (See Tables 7-1 and 7-3.)

Thread-Chasing Dial

In order to cut, or "chase," a complete thread on a lathe with a single-point tool, a number of cuts is required, each successive cut enlarging the thread groove until it has reached the depth required. The tool is withdrawn at the end of each cut and returned to the start of the thread at the tailstock end of the work. When starting the next cut the tool must enter the thread groove made by the first cut. One way to do this is to leave the split nut engaged and reverse the lathe in order to return the cutting tool. This method is objectionable because it is slow. It is better to disengage the split nut at the end of each cut and return the carriage by hand to its starting point. (Some large lathes have a power rapid traverse for moving the carriage.) When the split nut is disengaged from the lead screw, the tool must be made to enter the thread groove that is being cut when the carriage is again engaged with the lead screw. This is the function performed by the thread-chasing dial, Fig. 13-10.

The thread-chasing dial gives a visual indication as to when the split nut lever should be actuated to engage the carriage to the lead screw, in order to make the tool enter the thread groove. It is attached to the carriage and has a worm gear that engages the lead screw. The worm gear will revolve when the lead screw is turning and the carriage is stationary, or when the carriage is moved longitudinally, with the split nut disengaged and the lead screw stationary. When the carriage is moved by the lead screw, the worm gear will not revolve. A graduated dial is connected to the worm gear by a vertical shaft. The graduated dial is used by engaging the lead screw when one of the graduations is opposite the reference mark.

The thread-chasing dial operates on the following principle: The number of teeth on the worm gear of the thread-chasing dial is some multiple of the number of threads per inch of the lead screw. The number of teeth on the worm gear divided by the threads per inch of the lead screw will equal the number of graduations on the dial. For example, if the worm gear has 24 teeth and the lead screw has 6 threads per inch, the dial would have four divisions—each representing an inch of carriage travel. Then, if the carriage should be moved one inch with the lead screw stationary, six teeth on the worm would be engaged by the six threads on the

Fig. 13-10. Thread-chasing dial.

lead screw, thereby causing the worm and the connected dial to turn one-quarter of a turn (6 teeth ÷ 24 teeth = ¼ turn) which is equivalent to one mark or graduation on the dial since there are four equally spaced graduations around the dial. By subdividing the dial into eight equally spaced graduations, each line will correspond to a carriage movement of ½ inch. The dial, therefore, will enable the carriage to be engaged with the lead screw at points equal to a travel of one-half inch.

To illustrate the significance of understanding the principle of the thread-chasing dial, assume 13 threads per inch are being cut and that the work is stationary with the point of the tool in a thread. The lead screw is disengaged and the carriage is moved ½ inch, or one mark on the dial. The point of the tool will be in the middle—between the sixth and seventh thread. It will *not* be in the thread *groove*. If the carriage were now moved two marks on the dial, or one inch, the tool could enter a thread groove again. Thus, when the number of threads to be cut is odd, the carriage can only be engaged when one of the four graduations, representing an inch of travel, is opposite the reference mark. Suppose that ten threads per inch are being cut, with the work stationary and the point of the tool in a thread. When the lead screw is disengaged, the carriage is moved ½ inch and the tool will be in a thread groove since there are 5 even threads

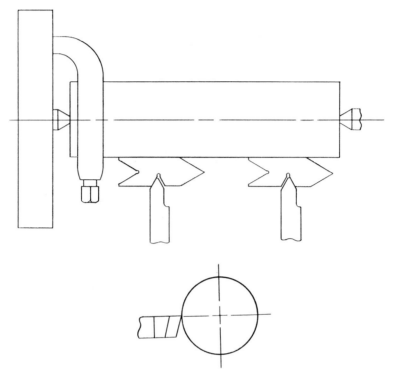

Fig. 13-11. Setting a thread-cutting tool with a center gage for cutting an American Standard Unified Thread Form.

in the ½-inch length of thread. Obviously, the tool will be in a thread groove when the carriage has moved one inch. Thus, it can be concluded that the carriage can be engaged when any mark on the dial comes opposite the index. If 11½ threads per inch are to be cut, the dial must be engaged on graduations representing a two-inch movement of the carriage, because in this instance, there will be a whole number of threads — in this case 23. Thus, the split nut must be engaged at every fourth graduation. Since every second graduation on the dial is numbered, this would mean that the split nut can be engaged at numbers 1 and 3, or 2 and 4 when cutting threads such as 11½, 12½, 15½, etc.

There are differences in design in thread-chasing dials; however, in principle they are all the same. To determine how to use a particular dial, simply move the carriage, longitudinally, a distance of ½ inch with the spindle stationary and note if the dial has turned one mark. Next, move the carriage one inch and note which marks on the dial represent the one-inch movement. Remember that *even* number threads per inch are cut by engaging the carriage at the ½-inch mark; *odd* number threads per inch are cut by engaging the carriage at the one-inch mark; and *fractional pitches*, such as 11½ threads per inch, are cut by engaging the carriage at two-inch marks.

Setting the Thread-Cutting Tool

The thread-cutting tool must be set accurately, otherwise the form of the thread that is cut will be incorrect even if the tool is correctly shaped. It should be set at the height of the lathe centers. The angular relationship of the thread is obtained by holding a center gage against the work, or against any convenient cylindrical surface that is parallel to the axis of the lathe, and comparing the sides of the threading tool with the sides of the 60-degree notch in the side of the center gage as shown in Fig. 13-11. The tool should not be jammed tightly in the 60-degree notch. Enough space should be left free to allow the center gage to be moved sideways so that each side can be checked independently of the other. It will be helpful to place a piece of white paper that will reflect light underneath the center gage. Thread-boring tools are set in a similar fashion as is seen in Fig. 13-16. Acme thread-cutting tools are also set in a similar manner except that an Acme thread gage is used in place of the center gage. For ultra-precision work, especially on smaller threads, the tool setting is sometimes given a final inspection by cutting a thread in a piece of brass with the tool. The thread is then inspected on an optical comparator or under a toolmakers microscope. Any errors in the position of the cutting tool can be seen on the thread and corrected. The procedure is repeated until the thread form is correct. For most ordinary classes of work using the center gage, or the thread gage, as described, is sufficient when done with care.

Cutting the Thread

Cutting the thread on the lathe, also called thread chasing, is illustrated in Fig. 13-12. The procedure for cutting the thread is shown in Figs. 13-13 and 13-14. In Fig. 13-13, the arrows show the direction in which the tool is moving and the letters have the following meaning: L.S., feed by lead screw; C.S., feed by cross slide; M.L., manual longitudinal feed; C.R., feed by compound rest slide. To begin with, start the

Fig. 13-12. Cutting American Standard Unified thread on a lathe.

290 CUTTING SCREW THREADS ON A LATHE Ch. 13

Fig. 13-13. Procedure for cutting an American Standard Unified thread. Arrows indicate direction of motion: L.S. Feed by lead screw; C.S. Feed by cross slide; M.L. Manual longitudinal feed; C.R. Compound-rest slide feed.

lathe spindle and move the tool toward the work until it just touches the surface to be threaded. Move the tool to clear the end of the work and then set the tool to take a cut of .002- to .005-inch depth. The tool is now in position 1, Fig. 13-13. At this point, the micrometer cross-feed dial is adjusted to read zero, without disturbing the position of the cross slide. On some lathes this cannot be done and the reading of this dial must be memorized. This position is called the *zero reference position* of the cross slide. When the proper mark on the thread-chasing dial is opposite the reference mark, the split-nut lever is actuated with a rapid motion to engage the carriage to the lead screw. The lead screw feeds the carriage longitudinally, causing the tool to cut a very shallow, winding groove on

Fig. 13-14. Schematic drawing showing how succeeding cuts (starting from the left) cut the thread from start to finish.

the work. This is represented by position 2, in Fig. 13-13. When the tool has traveled as far as the thread is wanted, as in position 3, it is withdrawn from the work by a quick turn of the cross-feed handle made with the left hand, while almost simultaneously actuating the split-nut lever with the right hand, to disengage the carriage from the lead screw. The timing of these two moves should be such that the cutting tool is just clear of the thread groove when the split-nut lever is actuated. The cutting tool is brought to position 4 by these moves, from which it is moved to position 5 using the manual, longitudinal feed which is actuated by turning the large carriage handwheel. At position 5, the cutting tool is set to take the next cut by feeding it to take a deeper cut with the compound rest. The amount of feed per pass through the thread groove can be accurately controlled by using the micrometer dial on the compound-rest feed screw. The cutting tool is then returned to the starting position, or position 1, by moving the cross slide until the micrometer cross-feed dial reads zero, or the reading registered when the tool was previously in this position. The cross slide is now at the zero reference position and the lathe is ready to take the next cut through the thread. Observe that the tool is fed into the work with the compound rest, and that the cross slide is always returned to the zero reference position for each cut through the thread groove. Before the second cut is taken the spindle should be stopped and the number of threads per inch cut by the first cut should be counted. If the lathe is not set up correctly the workpiece will not be ruined at this stage because of the shallow depth of the first cut. When the lathe is set up to cut the required threads per inch the spindle is started. The thread is cut to completion by taking successive cuts through the thread groove, repeating the manipulations used to cut the first groove.

Figure 13-14 is a schematic representation of the successive cuts used to cut the threads. No schedule of cuts that will fit each case can be given. In general, however, the first cuts are fairly heavy. Depending on the size of the thread being cut, and the class of work, the depth of the first cuts can be as little as .005 inch or as much as .060 inch. As an example, to cut a $\frac{1}{2}$-13 UNC-2A thread the initial cuts following the first "scratch" cut, which is made only on the first piece, might be .010- to .015-inch deep. The depth of cut per pass through the thread groove is then reduced until the final cuts used to finish the thread are only .001- to .0005-inch deep.

The size of the thread is not usually measured directly when it is in the process of being turned in a lathe. The procedure of bringing it to size is a cut-and-try procedure. The appearance of the thread will give an indication when it is getting close to the finish size. When it is judged to be close to the finish size, it is gaged with the ring gage, snap gage, or mating part, after each cut, until the gage fits properly as described in Chapter 12. Essentially the same procedure is used when measuring the threads by the three-wire method or with thread micrometers.

On some engine lathes the longitudinal-feed handwheel on the carriage can be disengaged. This is done to avoid the influence of the unbalanced force on this handwheel as it rotates. This force can cause the carriage

to feed unevenly when taking very light cuts such as when cutting threads. The uneven motion of the carriage results in errors appearing in the lead of the thread that is being cut; therefore, where possible, the handwheel should be disengaged when the carriage is engaged to the lead screw.

When cutting threads in steel and similar metals, a good grade of cutting fluid should be used to obtain a good surface finish on the threads. There are a good many proprietary cutting fluids. Generally lard oil or a heavy, sulfurized mineral oil will give good results. Threads should be cut in gray cast-iron without using any cutting fluid.

The infeed of the cutting tool for setting the tool to the desired depth of cut is made with the compound rest. Since the compound rest is set to 29 or 30 degrees, the tool moves toward the work at this angle. When the cut is taken the chip is therefore formed by the left side-cutting edge (ignoring, for the time being, the nose of the tool). Thus, only one chip is formed which is free to escape over the face of the tool. If, instead, the infeed of the tool was made by the cross slide, thereby feeding the tool straight in, two chips would be formed. One chip would be formed by the left side-cutting edge and the other by the right side-cutting edge. The two chips would flow toward each other and collide. This crowding of the chip causes the surface of the thread to tear and produces a poor surface finish.

When the infeed of the tool is made with the compound rest set at 29 degrees, the right side-cutting edge rubs against, or burnishes, the "back" side of the thread, thereby improving its surface finish. Regardless of the angle of infeed, the "back" side of the thread is sometimes rough and must be cut with the right side-cutting edge of the tool in order to "clean up." The procedure for "cleaning up" the back side of the thread is shown schematically in Fig. 13-15. This must be done before the thread is to size because it is necessary to take several cuts through the thread groove, thereby enlarging it. Position 1, Fig. 13-15 shows the incompleted thread groove which has been cut using a 29- or 30-degree infeed of the tool. The tool is moved to cut with the right side-cutting edge by backing out

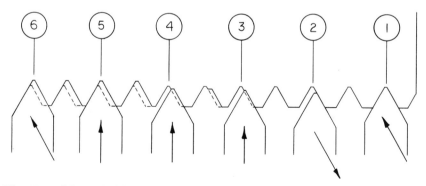

Fig. 13-15. Schematic illustration of procedure for "cleaning up" the back side of a thread.

the compound rest a distance of approximately .010 to .015 inch. Use the micrometer compound-rest feed dial to ascertain this distance and be sure to allow for the lost motion in the compound-rest feed screw. This is shown schematically in position 2, which shows the effect of this movement in relation to the thread. Actually, the movement is made before the cut is started, when the tool is clear of the thread. For the next several cuts the infeed of the tool is made with the cross slide which must be moved past the zero reference position. Positions 3, 4, and 5 show how the right side-cutting edge cuts the "back side" of the thread as a result of feeding with the cross slide. When the left side-cutting edge of the tool again touches the "front side" of the thread, as shown at position 5, a new zero reference position for the cross slide is obtained by adjusting the micrometer cross-feed dial to read zero or by memorizing the reading. The thread is then finished to size by infeeding the cutting tool with the compound rest as before (position 6).

It is sometimes necessary to align a thread-cutting tool with an existing thread groove. For example, this must be done if the tool has to be replaced before the thread is finished; or when the thread is rough cut with a roughing tool and finished with a finishing tool which is sometimes the procedure used for cutting large threads. The new tool must be set to a center gage and to the height of the centers as before. It is aligned with the thread by engaging the lead screw and turning the lathe forward, until the carriage has moved the tool opposite the thread grooves. The spindle is then stopped, but the lead screw is kept in engagement. This motion of the carriage takes up any lost motion in both the gearing and the split nut. The tool is then positioned to fit into a thread groove by moving the cross slide and the compound rest slide until the threading tool, as determined visually, is positioned to a perfect fit in the thread groove. The movement of the compound rest slide, being at an angle, moves the tool both longitudinally and perpendicularly to the lathe axis. Errors caused by the perpendicular movement of the compound rest can be corrected by the movement of the cross slide until the tool fits in the thread groove. When the tool is accurately positioned in the thread groove, the reading of the cross-feed dial determines the new zero reference position for cutting the thread, or it can be set to read zero. The tool is then withdrawn from the thread. The carriage is disengaged from the lead screw and again is brought to the starting position (position 1, Fig. 13-13) to cut the thread.

Cutting Left-Hand Threads

Left-hand threads are cut by reversing the direction of rotation of the lead screw so that the carriage will move from left to right. The compound rest is positioned at 29 or 30 degrees; however, the position is not the same as that for right-hand threads. It is positioned so that the right side-cutting edge of the threading tool will form the chip. The left-hand thread-cutting tool should have a 10- to 17-degree relief angle on the flank below the right side-cutting edge, and an 8- to 12-degree relief angle

below the left-hand cutting edge. The procedure for cutting the left-hand thread is essentially the same as for cutting the right-hand thread except for the differences explained below. If possible, the part with the left-hand thread should be designed with a neck, or groove, ahead of the cut. This provides a space in which the tool can be positioned in starting the cut. If a neck cannot be cut ahead of the thread, the best procedure is to start cutting the thread manually. The zero reference position of the cross slide is established in the same manner as for right-hand threads. The infeed of the tool is made with the compound-rest slide before the start of each cut. The cut is started by turning the spindle by hand, with the carriage engaged to the lead screw. When the tool approaches the start of the thread it is carefully moved into the thread groove, with the cross slide. The cross slide is moved in until the micrometer dial reaches the zero reference position, while at the same time slowly continuing to turn the spindle manually. When the zero reference position has been reached, the spindle is then started, and the remainder of the thread is finish cut by power.

Boring Internal Threads

Internal threads can be cut with a single-point tool held on the end of a boring bar and using a procedure that is very similar to turning external threads. The thread-boring tool is shaped to conform to the thread that is being bored. The relief angles on the thread-boring tool must be made large enough to clear the bore and are, therefore, dependent on the diameter of the bore. The part to be bored may be held in a lathe chuck or mounted on the faceplate. The hole to be bored must first be bored to the minor diameter of the thread. The lathe is set up to cut the required number of threads per inch. The compound rest is positioned at 30 degrees, with respect to the cross slide, as shown in Fig. 13-16. It may also be set at 180 degrees from the position shown; however, in some cases this position will cause the compound rest to come in contact with the work or the chuck. The position shown in Fig. 13-16 is usually preferred. The thread-boring tool should be set at the height of the lathe centers. A surface gage can be used to compare the height of the boring tool with the tailstock center. The thread-boring tool is set in correct relationship to the work with a center gage, as shown.

The procedure for boring the hole is the same as for cutting an external thread. Using the micrometer cross-feed dial as a reference, the cross slide is set at a zero reference position. After each cut, the cross slide is moved *forward* to clear the tool. After the tool is withdrawn from the bore it is returned to the zero reference position, as established by the cross-feed dial, in preparation for the following cut. The infeed of the tool to set the depth of cut for each pass is made with the compound-rest slide. The size of the thread is determined by the fit of a thread plug-gage or by a mating part. Small-diameter holes that have threads cut deep into the hole are sometimes finished by running a tap through the hole. The shank end of the tap is held in the tailstock center and the tap is turned by a

Fig. 13-16. Setup for boring thread and method of setting thread-boring tool.

wrench, with the lathe spindle held stationary. With this procedure the threads are uniform throughout the bore.

It is sometimes difficult to see the end of the cut when boring threads in a hole, and it is difficult to tell exactly when to actuate the split-nut lever to disengage the carriage from the lead screw. This becomes unnecessary if the longitudinal feed can be used to feed the tool out of the hole, rather than into the hole, in order to cut the thread. Left-hand threads can be bored in this manner with the setup shown in Fig. 13-16; however, right-hand threads must be bored feeding *into* the hole with this setup. On lathes equipped with a long taper, or cam-lock spindle nose, right-hand threads can be cut with the longitudinal feed feeding the tool out of the hole by cutting the threads with the lathe spindle turning in the reverse, or clockwise, direction. When this procedure is used, the lathe must be set up as shown in Fig. 13-17, in order to bore the thread. The thread is actually cut on the "back side" of the bore, as shown. The thread-

Fig. 13-17. Alternate method of boring a right-hand thread on lathe equipped with tapered, or cam-lock, spindle nose.

cutting tool is positioned in the internal groove to start the cut with the cross slide in the zero reference position, as determined by the cross-feed dial. With the lathe spindle running in reverse, the split nut is engaged at the proper time (determined by the threading dial), and the carriage is moved from left to right to bring the tool out of the bore. The infeed of the tool is made with the compound rest, and the cross slide is backed off to allow the tool to clear the threads as it reenters the hole. The cross slide is then positioned to the zero reference position in preparation for the next cut. This procedure is repeated until the thread is cut to size. Since the tool is clear of the work at the end of each cut, no special care is required when actuating the split-nut lever to disengage the cross slide from the feed screw. Higher spindle speeds may be used because there is less danger of the threading tool feeding too far into the bore. This procedure is not recommended for lathes having a threaded spindle nose be-

cause of the possibility of the chuck or faceplate unscrewing itself from the spindle. Likewise, boring bars with threaded ends cannot be used.

Cutting Acme Threads

The general procedure for cutting an Acme thread is the same as for cutting a 60-degree American Standard Unified and National thread. An Acme threading tool with a 29-degree included angle must be used. It is set with an Acme thread gage after the compound rest has been positioned at $14\frac{1}{2}$ degrees, with respect to the cross slide. The micrometer cross-feed dial is used to establish the zero reference position of the cross slide and the cross slide is returned to this position before starting each cut. The infeed of the cutting tool, in order to obtain the depth for each cut across the work, is made with the compound rest so that the tool will cut with only two cutting edges, namely, the right side-cutting edge and the end-cutting edge. If required, the "back side" of the thread can be cleaned up using the same procedure previously described. Many Acme threads are very long, as shown in Fig. 13-18, and the work must be given support to prevent it from being deflected by the cutting forces. A steady rest cannot be used when cutting the thread as in the illustration, because it would interfere with the movement of the carriage; therefore, a follower rest must be used to support the work.

Taper Threads

Taper threads should be cut with a taper attachment if at all possible. When the required taper is obtained by setting the tailstock off-center, the thread will not advance at a uniform rate, nor form a true helix. This

Courtesy of the Clausing Division of the Atlas Press Company

Fig. 13-18. Cutting an Acme thread for a lead screw.

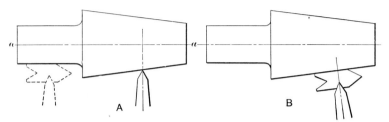

Fig. 13-19. Correct and incorrect positions of the tool for taper thread cutting

"drunken thread," or error, is caused by the angularity between the driving dog and the driven plate, which causes the rotating speed of the work to vary during each revolution. The bearing surface between the lathe centers and the work centers (when the tailstock is offset) is another cause of inaccuracy, because as the work centers wear rapidly due to the poor bearing surfaces, the angle of the taper is changed as the tailstock spindle is tightened. The amount of the error depends upon the angle of the taper and the distance that the centers must be offset.

The cutting tool must be set having the included angle of the thread bisected by a line that is perpendicular to the axis of the work, as shown at A, Fig. 13-19. It cannot be set by holding the center gage against the tapered surface as shown at B, which would cause the form of the thread that is cut to be incorrect. The taper attachment can be adjusted to the correct taper by placing a taper gage (not threaded) of the required taper between centers. A dial test indicator is fastened in the toolholder and the contact point of the indicator is placed against the gage at the height of the lathe centers. The taper attachment is adjusted until the hand of the indicator does not move when it is traversed from one end of the gage to the other.

Cutting Multiple Threads

The general procedure for cutting multiple threads is the same as for cutting single threads. Figure 13-20 shows a multiple thread and the method used to cut the thread. The first thread is cut as though a single thread were being cut, leaving the work as shown at A. When this cut is finished the work must be indexed one-half of a revolution (for a double thread) without disturbing the position of the lead screw or carriage. This will place the tool in a position to cut the second thread groove, centrally, between the grooves of the first thread, as indicated by the tool shown in dotted lines. The second groove is then cut producing a double thread, as shown at B.

The lead, which is the distance that the thread will advance in one turn, is not equal to the pitch—for multiple threads. For example, in Fig. 13-20, the pitch of the double square thread is $\frac{1}{4}$ inch and the lead is $\frac{1}{2}$ inch. If this had been a triple square thread the lead would have been $\frac{3}{4}$ inch. The cutting tool must travel a distance equal to the lead per revo-

Ch. 13 CUTTING SCREW THREADS ON A LATHE 299

lution of the work. In the case of multiple threads, the lathe must be set to cut an equivalent number of threads per inch, which is dependent upon the lead of the thread, and which is not equal to the actual number of threads per inch. The equivalent number of threads per inch can be calculated by the following formula:

$$\text{Equivalent threads per inch} = \frac{1}{\text{Lead}} \qquad (13\text{-}4)$$

For the threads in Fig. 13-20 the equivalent threads per inch would be:

$$\text{Equivalent threads per inch} = \frac{1}{\text{Lead}} = \frac{1}{\frac{1}{2}} = 2$$

The actual number of threads per inch in this case is four.

The indexing, or spacing, of multiple threads can be done in several different ways. One method is to disengage the gear on the stud (Figs. 13-9 and 6-4) from the idler, or quadrant, gear and turn the stud shaft a predetermined amount while the idler, or quadrant, gears remain stationary. The stud gear is then reassembled to the stud, in engagement with the idler, or quadrant, gear. The stud shaft is turned the predetermined amount rotating the stud gear a given number of teeth. Before starting to turn the stud shaft a stud-gear tooth and the mating tooth-space on the idler, or quadrant, gear are marked with chalk. If the number of teeth to

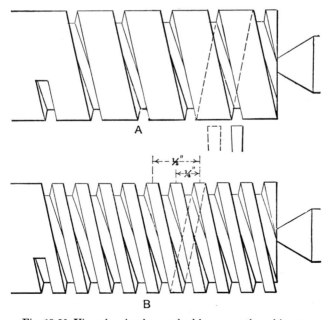

Fig. 13-20. View showing how a double square-thread is cut.

be indexed is 16, then the sixteenth tooth from the marked tooth on the stud is placed in the marked tooth space in the idler, or quadrant, gear.

The number of teeth on the stud gear that must be indexed is determined as follows:

1. Count the number of teeth through which the stud gear is turned when the spindle is turned one revolution.
2. Divide this number by the number of threads that are to be cut.
3. The answer must be a whole number, or the stud gear must be changed.

Example 13-5:

A double thread is to be cut on a lathe having a 24-tooth stud gear.

When the spindle is turned one revolution it is found that the stud gear turned through 18 teeth.

The number of teeth which the stud gear must be indexed is:

$$\frac{18}{2} = 9 \text{ teeth}$$

Example 13-6:

A triple thread is to be cut on a lathe having a 24-tooth stud gear.

When the spindle is turned one revolution it is found that the stud gear turned through 16 teeth.

The number of teeth which the stud gear must be indexed is:

$$\frac{16}{3} = 5\frac{1}{3}$$

This, however, is not the answer since it is *not* a whole number.

When the answer is not a whole number use the following procedure, illustrated by solving the problem in Example 13-6, above:

1. Count the number of teeth through which the stud gear is turned, per revolution of the spindle, which in this case is 16.
2. Calculate the stud ratio or the ratio between the spindle and the stud shaft.

$$\text{Stud ratio} = \frac{\text{Number of teeth per revolution of spindle}}{\text{Number of teeth on stud gear}}$$

$$= \frac{16}{24} = \frac{2}{3}$$

3. Expand the stud ratio until a numerator is found that is divisible by three, and a denominator that is equal to the number of teeth on a stud gear that is available.

$$\frac{2}{3} = \frac{2 \times 9}{3 \times 9} = \frac{18}{27}$$

Thus, the triple thread can be cut by using a 27-tooth stud gear and indexing this gear through 18 teeth.

If the lathe is not equipped with additional gears for the stud, a special gear must be made or another method of indexing the work must be used.

One method that can sometimes be used to obtain the spacing for multiple threads is to engage the carriage and the lead screw and move it forward just far enough to eliminate the lost motion in the split nut and the gearing. Attach a dial test indicator to the lathe bed of the carriage and place a stack of precision-gage blocks between the indicator and the carriage. The length of the stack of gage blocks is made equal to the pitch (not the lead) of the thread. Zero the indicator to the gage blocks and then remove the blocks. Disengage the feed-thread change handle or the quadrant gear (Fig. 6-4) to disconnect the lead screw from the spindle. Turn the lead screw manually, with the carriage engaged, until it comes against the dial test indicator making it read zero. Reengage the feed-thread change handle or the quadrant gear. The carriage can also be moved the required distance, using the micrometer carriage positioning attachment, Fig. 6-19. When multiple threads must be cut frequently, a special indexing faceplate, or a special tool block that accurately spaces the cutting tool can be used.

The thread-chasing dial can also be used to space double threads whenever the equivalent threads per inch are an odd numbered thread as, for example, 5, 7, 11, etc. The first thread is cut using the graduations on the thread-chasing dial that represent one-inch movements of the carriage, while the second thread is cut by the alternate graduations which represent ½-inch movements of the carriage. If the equivalent threads per inch end in a "½," such as 5½, 7½, 11½, etc., the first thread is cut by any one graduation, say, number 1, and the second thread is cut by the graduation that represents a two-inch movement of the carriage. In the case of the dial shown in Fig. 13-10, this would be graduation number 3. Graduations 2 and 4 could also be used. The use of the thread-chasing dial as a means of spacing double threads is recommended whenever possible. The procedure to use when the thread dial is employed to space the threads, is to alternately cut one thread groove and then the other so that both thread grooves are finished at nearly the same time.

QUESTIONS

13-1. Name the constructional features of an engine lathe that are designed primarily to cut screw threads.

13-2. What rake angle should be ground on thread-cutting tools?

13-3. What is the advantage of using an offset thread-cutting tool?

13-4. How is the form of thread-cutting tools inspected?

13-5. What would be the width of the end of an Acme tool when the vertical vernier-scale of a gear-tooth vernier is set at .250 inch, the caliper-jaw vernier scale reads .4793-inch, and the jaws touch the sides of the thread tool?

13-6. How does a lathe cut 8 threads per inch, assuming that it is set up to cut this thread?

13-7. Calculate the simple gears that can be used to cut 16 threads per inch on a lathe having a lead number equal to 6.

13-8. Calculate compound gears required to cut a thread having 14 threads per inch, on a lathe, when the lead number is 8.

13-9. Calculate compound gears required to cut a thread having 5 threads per inch, on a lathe, when the lead number is 6.

13-10. Calculate the gears required to cut a thread having a pitch of 2.40 millimeters, on a lathe having a lead number of 6.

13-11. Calculate the gears required to cut a worm thread for an 8-diametral-pitch worm gear. The lead number of the lathe is 6.

13-12. How is the spindle speed for cutting screw threads determined? Name all of the factors that must be considered.

13-13. Describe the construction of a thread-chasing dial.

13-14. A certain lathe lead-screw has 4 threads per inch. The worm gear on the thread-chasing dial has 24 teeth. How many equally spaced graduations are required on the face of the dial so that each graduation represents a movement of the carriage of one inch? Of $\frac{1}{2}$ inch?

13-15. Refer to Fig. 13-10. Name the graduations on the thread-chasing dial that can be used to cut the following threads:

4 threads per inch	32 threads per inch
7 threads per inch	21 threads per inch
12 threads per inch	8½ threads per inch

13-16. How is the thread-cutting tool set for cutting an American Standard Unified thread?

13-17. Describe the procedure used for cutting an American Standard Unified thread on an engine lathe.

13-18. Why is the infeed of the cutting tool for setting the depth of cut made with the compound rest?

13-19. How is the "back" side of the thread cleaned up?

13-20. What is the difference between cutting right- and left-hand threads?

13-21. What is the advantage of boring right-hand screw threads with the spindle rotation reversed?

13-22. The lead of a triple thread is equal to .250 inch. What are the equivalent threads per inch?

13-23. A triple thread is to be cut on a lathe having a 24-tooth stud gear. When the lathe spindle is turned one revolution, the stud gear turns through 18 teeth. Determine the number of teeth that the stud gear must be indexed to space the three thread-grooves on the work.

13-24. Assume that the thread in Problem 13-23, above, is a quadruple thread instead of a triple thread. Calculate a gear and the amount that it should be indexed in order to cut this thread.

13-25. A double-thread screw having an actual 13 threads per inch is to be cut on an engine having a thread-chasing dial like the one in Fig. 13-10. Describe how to cut this thread by using the thread-chasing dial.

CHAPTER **14**

Turret Lathes, Production Lathes, and Vertical Lathes

While engine lathes are very versatile and are easily adapted to perform a wide range of different operations, generally, they do not meet the requirements demanded of modern turning equipment in producing large quantities of duplicate parts at high production rates. This need is met by turret lathes, automatic lathes, and vertical turret lathes. These machine tools are designed to take cuts with several cutting tools, simultaneously. Each cutting tool is pre-positioned, thereby reducing or eliminating entirely the need for setting each tool for the cut and for taking trial cuts to obtain the required size on the part. Some of these lathes operate on an automatic cycle, while others are manipulated by an operator.

Some vertical lathes are not designed for high production but for performing turning and boring operations on parts that are more conveniently held on a faceplate or chuck. These have a *horizontal surface* as compared to the *vertical holding surfaces* on other lathes. This type of machine is commonly referred to as a vertical boring mill. They are often used to bore long, deep holes because the long boring bar is held in a vertical position. If these boring bars were held in a horizontal position they would bend or sag appreciably due to their own weight. Vertical lathes can also be tooled for the quantity production of duplicate parts. Many modern vertical lathes have been equipped with numerical controls which enable them to turn and bore complex contoured surfaces in addition to performing other operations.

Turret Lathes

The characteristic feature of a turret lathe is the turret which is located at the rear of the bed in the position occupied by the tailstock on engine lathes. The turret has six faces, on each of which a variety of cutting tools can be mounted. The turret can be indexed to bring the cutting tools on one of the turret faces into a position to perform cutting operations on the part, while the tools on the other turrets are held conveniently out of the way. The turret lathe also has a carriage which, like the engine-lathe carriage, provides a cross-feed and a longitudinal-feed motion to the tools that are mounted on it. Turret lathes usually have two aprons, one for the carriage and the other for the saddle upon which the

Ch. 14 TURRET, PRODUCTION, AND VERTICAL LATHES

turret is mounted. Both of these aprons contain selective gear trains which provide the available feed rates. The necessary handwheels, handles, and levers for actuating these feeds are mounted outside of the apron. Normally, turret lathes do not have a compound rest, although one can be mounted on the cross slide, in some cases, as an attachment. The headstock contains the spindle and the gears for making the spindle turn at the selected revolutions per minute.

There are two basic types of turret lathe constructions, the ram type shown in Fig. 14-1 and the saddle type shown in Fig. 14-2. The hexagon turret of the ram type turret lathe is mounted on a ram slide which can be rapidly moved toward or away from the spindle, by the turnstile. It can also be fed by an automatic power feed toward the spindle. Stop screws provide a positive stop and an automatic disengagement of the longitudinal feed. When the saddle is brought back, the turret is automatically indexed to bring the next face into the operating position. The ram slide moves easily and can be quickly handled by the operator. It slides on a saddle that can be fastened to the lathe bed at different distances from the spindle. When the turret lathe is operating, however, the saddle is fastened to the bed and remains stationary. The carriage is provided with both forward and reverse longitudinal and cross feeds. The longitudinal feed is provided with a positive stop which also disengages the automatic feed. The cross slide has trip dogs that disengage the automatic cross feed; however, these dogs can be passed over manually and the tool can be positioned for taking a cut by reading the graduations on the micrometer cross-feed dial. The cross slide is a bridge type which, like the cross slide on an engine lathe, bridges the opening in the lathe bed. It is finish machined on the top and sides. The top has a T-slot which provides an anchor for the bolts used to clamp tool blocks in place. Two standard tool blocks that satisfy many of the requirements are the rear tool post for cutting from the back side of the cross slide and the square turret whch is mounted in the front. The square turret can hold four different cutting tools and it can be indexed to bring the appropriate cutting tool into position for taking a cut. The ram type turret lathes are usually smaller than the saddle type turret lathes and are generally used to machine smaller parts.

The saddle type turret lathe, Fig. 14-2, is usually a heavier and larger machine that is used to machine comparatively large parts. On the saddle type turret lathe the turret is mounted directly on the saddle which is moved longitudinally back and forth by the handwheel. It can also be moved back and forth by means of a power rapid-traverse. Automatic power feeds are used to feed the saddle toward the headstock when taking cuts. A stop roll, located inside of the bed, provides a positive stop and disengages the feed of the saddle. It can be passed over by swinging a master stop out of position. The turret on saddle type turret lathes must be indexed by hand. It is rigidly supported by the saddle at all times, in contrast to the ram on ram type turret lathes, which may well overhang the saddle a considerable distance when taking cuts. For the lighter work

Fig. 14-1. Principal parts of a ram type turret lathe.

Ch. 14 TURRET, PRODUCTION, AND VERTICAL LATHES 307

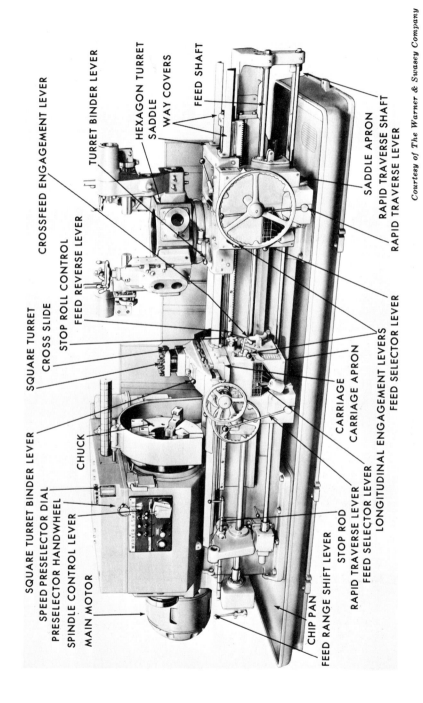

Fig. 14-2. Principal parts of a saddle type turret lathe.

generally done on ram type turret lathes this is usually no disadvantage whereas on saddle type turret lathes, which usually take heavier cuts, the rigid support of the saddle is a very decided advantage. Some saddle type turret lathes are equipped with a cross-sliding saddle. This provides a cross-feed direction to the entire turret which can be used for internal recessing and facing operations. Standard turning and boring heads are shown mounted on the faces of the turret in Fig. 14-2. When these are removed, the face of the turret appears the same as in Fig. 14-1. The carriage of the saddle type turret lathe, Fig. 14-2, is generally a side-mounted or "side-hung" type, although a bridge type cross slide can be provided. The bridge, however, places a limit on the diameter of the work that can be placed over it. The advantage of the side-mounted carriage is that the cutting tool held on this carriage is rigidly supported when turning larger diameters. The side-mounted carriage is supported by one of the ways on the top of the bed and by the way that is machined on the front side of the bed. It can be moved in two directions longitudinally by the apron handwheel, by rapid traverse, and by automatic power feed. A stop rod which engages a micrometer stop-roll on the carriage provides an accurate means of positioning the carriage in the longitudinal direction. It will also automatically disengage the power longitudinal feed when the carriage contacts the stop roll. The automatic cross feed is disengaged by stop dogs which can be passed over by the manual cross feed. The micrometer cross-feed dial is used to accurately position the tool for establishing the depth of cut.

Turret lathes are also classified as bar and chucking machines. Bar type turret lathes are used to machine parts directly from raw bar stock. The bar is placed through the spindle of the lathe. One end of the bar is supported by a bar chuck that is held on a bar feed rack which is attached to the end of the headstock. The other end of the bar is held in a collet chuck which is located inside the spindle nose. When the collet chuck is opened, the bar stock is fed out of the spindle by the bar feed, against a positive stop located on one of the turret faces. The collet chuck is then closed and the part is made from that portion of the bar stock protruding from the nose of the spindle. As a final operation, it is cut off with a cut-off tool. Chucking type turret lathes are used to hold the work in a three- or four-jaw chuck, on a faceplate, or in a fixture that is held on the faceplate while performing the various turning and boring operations. Most turret lathes are designed to be readily converted from one type to the other. Ram type turret lathes, particularly, can be readily converted for either bar work or chucking work. The work done on ram type turret lathes is probably equally divided between bar work and chucking work. Saddle type turret lathes are more frequently used to do chucking work.

An automatic bar and chucking turret lathe is shown in Fig. 14-3. This machine can be set up to do either bar or chucking work. A machining program for the part to be machined is first made, after which the proper tooling is set up on each station. The automatic cycle is set up by adjust-

Ch. 14 TURRET, PRODUCTION, AND VERTICAL LATHES 309

Courtesy of the Gisholt Machine Company

Fig. 14-3. Automatic bar and chucking turret-lathe.

ing the traverse and feed-trip settings. The other machine functions are then set from one station located in front of the turret saddles, and the electrical function controls located on the panel at the front of the headstock. The final adjustments for obtaining the size of the part are made when machining the first part. After it has been set up the automatic turret lathe will repeat the entire cycle of operations without further attention of the operator—except to load and unload parts held in the chuck, or to insert new bar stock through the spindle. The operator, or a set-up man, must replace dull cutting tools when required.

Turret-Lathe Tooling and Machining Methods

The turret lathe is capable of performing all of the operations that can be performed on an engine lathe. The method by which these operations are performed on the turret lathe sometimes differs from the method used on an engine lathe because of the tooling that is available for use on the turret lathe. Most of the turret-lathe tooling is standard tooling that can be used to do a variety of work. Some special tooling is also used, if required by the complexity of the workpiece, or if the quantities to be produced are large enough to warrant the extra expense necessary to obtain faster production.

Standard turret-lathe tools are classified as chucking tools, cross-slide tools, and bar tools. Chucking tools are held on the turret and are used

to machine parts that are held in three- or four-jaw chucks. Typical of the parts made in this way are large gear blanks, large collars, large bushings, pulleys and sheaves. Bar tools are used to make parts from bar stock such as small gear blanks, small bushings, shafts, studs, screws, and bolts. Cross-slide tools can be used for both classes of work. One additional category of tools should be mentioned and this is work-holding devices. These include collet chucks, three-jaw chucks, four-jaw chucks, faceplates, and special faceplate work-holding fixtures. The cutting tools are the same, or similar, to those used on engine lathes. The cutting speed is calculated the same as for engine-lathe work and the feed rates are also selected on the same basis. The possible ways in which these tools can be set up is almost without limit. Single-point cutting tools that are essentially the same as those described in Chapter 7 are used on the cross slide; therefore, they will not be treated in this chapter. Attention will be focused on some of the basic chucking and bar tools that are attached to the turret. These tools are introduced by referring to the appropriate illustration which should be studied along with the descriptions that follow.

Figure 14-4, Overhead Turning. This operation consists of turning one or more surfaces on the outside diameter of the workpiece with cutting tools that are positioned above the workpiece and are fed by the turret. The cutting tools are held by *cutter holders,* one of which is plain, while the other is adjustable. The cutter holders are held by a *multiple turning head* which is mounted on a face of the turret. Several cutter holders can be mounted on the multiple turning head. In Fig. 14-4, two cuts are taken on the outside diameter, while the end of the part is faced with a tool that is held by a square turret mounted on the cross slide.

Figure 14-5, Multiple Turning and Drilling. The multiple turning head can be used to turn, while simultaneously machining the inside of the hole

Courtesy of The Warner & Swasey Company

Fig. 14-4. (Left) Overhead turning. Fig. 14-5. (Right) Multiple turning and drilling.

Ch. 14 TURRET, PRODUCTION, AND VERTICAL LATHES 311

Courtesy of The Warner & Swasey Company

Fig. 14-6. (Left) Boring. Fig. 14-7. (Right) Start drilling.

with a drill, reamer, or boring bar. In the illustration two turning cuts are being taken simultaneously; one from the square turret and the other by tooling mounted on the turret face. The drill shown is a core drill, used to enlarge a cored hole in a casting. It has more cutting edges than a twist drill, thereby enabling a faster feed to be used; however, it cannot be used to open a hole from solid metal as the cutting edges do not come to a point.

Figure 14-6, Boring. Boring operations are frequently performed on turret lathes. One method is to hold the boring bar in a *slide tool* that is mounted on the face of the turret. The slide tool has a micrometer that is used to adjust the boring tool to obtain the required size of the bore quickly and accurately. Stops are provided on the slide tool so that the boring tool, when it is being withdrawn, can be backed away from the bore to prevent the tool from marking the finished surface. Boring operations can also be performed on turret lathes with the boring tool held by the square turret which is mounted on the cross slide.

Figure 14-7, Start Drilling. The purpose of the start drilling operation is to spot a true cone on the end of the workpiece which will act as a guide for the subsequent drilling operation. The start drill is short and stiff. It will not deflect when starting the hole, and the coned spot will be concentric with the rotation of the lathe. The drill which follows will be guided by this cone and will drill a hole which is also concentric. Although special start drills are frequently used, combination drill and countersinks, or center drills, are also used for this purpose.

Figure 14-8, Combination Stock-Stop and Start-Drill. The first operation when doing bar work is generally called stock stop. This operation consists of feeding the bar stock through the spindle until it contacts a stop that is held in the turret. This stop allows just the right length of bar stock to protrude from the face of the collet chuck in order to make the part. The collet chuck is closed when the bar stock comes against the stock

Courtesy of The Warner & Swasey Company

Fig. 14-8. (Left) Combination stock-stop and start-drill. Fig. 14-9. (Right) Spade drilling.

stop. The stock stop shown is combined with a start drill which is withdrawn during the stock-stop operation.

Figure 14-9, Spade Drilling. A spade drill is made from a flat piece of tool steel that is clamped to a bar-like spade-drill holder. The point of the spade drill is similar to the point of a twist drill providing it with the ability to open holes in solid metal. Spade drills are often used to drill large holes that are over 1¼ inches in diameter. Twist drills of this size are too long and overhang the turret excessively, whereas the overhang of spade drills is at a minimum. The blade of the spade drill can be replaced, when required, without the necessity of purchasing a new spade-drill holder. The setup for drilling a hole with a twist drill is similar to Fig. 14-9, and therefore will not be illustrated.

Courtesy of The Warner & Swasey Company

Fig. 14-10. (Left) Reaming. Fig. 14-11. (Right) Floating blade reamer.

Ch. 14 TURRET, PRODUCTION, AND VERTICAL LATHES 313

Figure 14-10, Reaming. Holes are finished by reaming from the turret, as shown. The reamer should be held in a *floating reamer holder* to assure proper alignment of the reamer with the drilled hole, thereby assuring that the reamed hole will be straight—not tapered.

Figure 14-11, Floating-Blade Reamers. Larger holes are often reamed with a floating-blade reamer. The two cutting blades of this reamer are allowed to float a certain amount, which provides the necessary alignment to obtain a straight hole.

Figure 14-12, Tapping. Tapping can be done on a turret lathe to produce internal threads. The tap must be held in a *releasing tap holder*. To tap the hole the turret is advanced toward the rotating workpiece, using sufficient force at the beginning in order to make the tap start to cut the threads. After the threads have been started the rotation of the workpiece and the lead of the threads will combine to pull the tap into the hole. As the tap feeds through the hole the turnstile is turned by hand just enough to keep the turret and the tap holder moving at the same rate as the tap. When the forward progress of the turret is stopped, the tap holder will release the tap, allowing it to revolve with the workpiece. The tap is withdrawn by reversing the direction of the spindle rotation.

Figure 14-13, External Threading. External threads are generally cut on turret lathes with self-opening die heads. As with tapping, the die head must be started to cut the threads. After the die has started, a light feeding force is used on the turnstile in order to keep the turret moving with the die head. When the forward progress of the turret is stopped the die head will snap open, thereby clearing the dies from the threads that have been cut. After the die head has been backed away from the threads it is closed by a movement of the handle in preparation for cutting the next thread. Die heads are available for cutting a wide range of thread sizes.

Courtesy of The Warner & Swasey Company

Fig. 14-12. (Left) Tapping. Fig. 14-13. (Right) External threading.

314 TURRET, PRODUCTION, AND VERTICAL LATHES Ch. 14

Courtesy of The Warner & Swasey Company
Fig. 14-14. (Left) Roller bar turner. Fig. 14-15. (Right) Cut-off operation.

Figure 14-14, Roller-Bar Turners. Roller-bar turners are used to turn long bar work which overhangs the collet. The rollers support the bar stock to keep it from deflecting. The rollers may be mounted ahead of the cutting tool in order to obtain concentricity between the previously turned diameter and the diameter being turned. The rollers are set behind the cutting tool, as shown, when turning the first diameter on rough-bar stock. When the rollers are set behind the cutting tool they burnish the turned surface thereby improving its appearance. At the end of the cut the cutting tool is withdrawn from the work by moving the cutter relieving lever. This prevents the cutting tool from marking the turned surface as the roller turner head is backed off the work by the turret.

Courtesy of The Warner & Swasey Company
Fig. 14-16. Tooling layout for machining a flange.

Ch. 14 TURRET, PRODUCTION, AND VERTICAL LATHES

Figure 14-15, Cutting Off. After a bar part has been completely machined it is cut off with a cut-off tool. A convenient place to mount this tool is in the rear cutter-block. When held by the rear cutter-block the cut-off tool must be held in an upside-down position. When the space is available, the cut-off tool can also be held by the square turret.

Most work done on a turret lathe is planned in advance. Often, tooling layouts are made which give the operator, or the set-up man, a pictorial representation of the way that the machine is to be set up. Figure 14-16 illustrates the tooling layout for machining a flange. The following sequence of operations is used to finish the part:

1. Start to true-bore hole G, deep enough to start core drill true.
2. Core drill hole G and rough-turn diameters A and D from turret. At the same time, rough-face F and C from square turret. Face F first and C last.
3. Bore hole G, to make sure that it is concentric.
4. Ream hole G to size.

Courtesy of The Warner & Swasey Company

Fig. 14-17. Ram type turret-lathe setup for bar work.

316 TURRET, PRODUCTION, AND VERTICAL LATHES Ch. 14

Courtesy of the Gisholt Machine Company

Fig. 14-18. Machining steel gate-valve bodies held in an indexing fixture on a saddle type turret lathe having a cross-sliding turret.

5. Finish-turn diameters A and D. At the same time, finish faces F and C.
6. Chamfer B and E.

Courtesy of the Gisholt Machine Company

Fig. 14-19. Automatic production lathe.

Ch. 14 TURRET, PRODUCTION, AND VERTICAL LATHES 317

Figure 14-17 shows a typical, ram type, turret-lathe setup for doing bar work. A saddle type turret lathe equipped with a cross-sliding turret is shown machining steel gate-valve bodies in Fig. 14-18. The gate valve is held in a special indexing fixture that permits three ends of the valve body to be machined in a single chucking. The versatility of the turret lathe for doing a wide variety of turning jobs is demonstrated by these illustrations.

Production Lathes

A typical automatic production lathe is shown in Fig. 14-19. These machines are used for high-volume repetitive work such as is encoun-

Courtesy of the Gisholt Machine Company

Fig. 14-20. Automatic production lathe tooled to produce automotive brake drum.

318 TURRET, PRODUCTION, AND VERTICAL LATHES Ch. 14

tered in the automotive industry. The tooling used on these machines is specially designed to do one particular job with great efficiency. A typical setup which is used to machine front and rear automotive brake drums is shown in Fig. 14-20. A special back-facing attachment is located inside the work-holding fixture for facing the mounting face of the brake drum. The front tool-slide turns and bores the other end of the drum. All turning and boring operations are completed on this machine in one setup. No great amount of skill is required to operate an automatic production lathe; however, the planning of the operations, the layout of the tooling, and the design of the tooling do require skill and experience of the highest order.

Vertical Lathes

The vertical lathe, commonly referred to as a *vertical boring mill*, has a circular table which revolves about a vertical axis so that the work-holding surface is horizontal, thus making it comparatively easy to place and hold large circular parts in position. (See Fig. 14-21.) The two large, vertical housings are connected together at the top by a bridge, to provide a rigid support. Ways are machined on the face of the housing on which the cross rail is mounted. The cross rail can be adjusted to any desired height above the top of the circular table. It is not, however, set in motion to take a cut. Two saddles, mounted on the cross rail, can be moved along the length of the cross rail. Tool bars, which have a vertical

Courtesy of The G. A. Gray Company

Fig. 14-21. Vertical boring mill.

Courtesy of The Bullard Company

Fig. 14-22. Vertical turret lathe machining jet-engine part.

movement, are mounted on each saddle. They can also be swiveled on the saddle to provide an angular movement. Tool blocks for holding cutting tools are located at the ends of the tool bar. The vertical orientation of the tool bars allows them to be made small enough to do boring operations without deflecting due to their own weight. Many vertical boring mills are equipped with a pentagon turret, in which case they are usually called *vertical turret lathes*. A vertical turret lathe is shown in Fig. 14-22 machining a large fan exit-case of a jet engine. The pentagon turret provides five tooling stations on each of which one or more cutting tools can be mounted. Both vertical boring mills and vertical turret lathes can be equipped with a side head which can be seen in Fig. 14-22. The side head can move in a vertical direction by sliding over the ways of the right-hand column. The side-head arm moves in a horizontal direction. The cutting tools mounted on the side head can, therefore, move parallel and perpendicular to the surface of the faceplate.

Workpieces can be bolted directly to the faceplate or they can be held by chuck jaws. The chuck jaws operate independently of each other

in a manner similar to four-jaw chucks. Figure 14-23 shows a cable hoist drum being machined on a vertical turret lathe. The workpiece is held by four-chuck jaws while simultaneous cuts are taken by tools held by the pentagon turret and the side head. Special work-holding fixtures, such as shown in Fig. 14-24, are sometimes used to hold the work. This workpiece is very thin and is apt to distort and vibrate when being cut. The fixture must also hold the part without distorting it. The part is firmly clamped to the base of the fixture. The snubbers at the top help to support the work against distortion due to the cutting forces, and they help to prevent vibration or chatter from occurring.

Vertical lathes are used to hold workpieces that can be more conveniently clamped to a horizontal surface rather than to a vertical surface. They are used to machine surfaces that require the use of a long, extended

Courtesy of the Gisholt Machine Company

Fig. 14-23. Cable hoist drum being machined on a vertical turret lathe.

Ch. 14 TURRET, PRODUCTION, AND VERTICAL LATHES 321

Courtesy of The Bullard Company

Fig. 14-24. Special work-holding fixture and tooling blocks on vertical turret lathe.

bar to hold a cutting tool that cannot be supported at two ends. The advantage of the vertical lathe in such situations is that the long bar—say, a boring bar—does not deflect under its own weight when it is hung in a vertical position. The vertical lathe is also often used in preference to other machine tools when large, internal, facing cuts are required; when very large diameter holes are to be bored; and when both large-diameter boring and facing cuts are required on the same part. Of course, cylindrical turning and facing cuts on outside surfaces can also be conveniently taken on these machines. Many vertical lathes are equipped with numerical controls or with tracer controls which add to their versatility. The vertical turret-lathe in Fig. 14-22 is equipped with tracer controls on the turret and on the side heads, enabling it to cut angles and complex contours. The machine in Fig. 14-23 is equipped with a numerical-control system.

QUESTIONS

14-1. What are the distinguishing features of a turret lathe as compared to an engine lathe?

14-2. What is the difference between a ram type and a saddle type turret lathe?

14-3. What is the difference between a bridge type and a side-mounted cross slide on a turret lathe?

14-4. What kind of work is done by a chucking machine?

14-5. What type of work is done by a bar machine?

14-6. What is a multiple turning-head and how is it used?

14-7. What is the purpose of the start drill?

14-8. What is the stock stop used for?

14-9. Describe three methods of turning cylindrical surfaces on turret lathes.

14-10. What is a spade drill and why is it used?

14-11. What is the advantage of a floating-blade reamer?

14-12. Describe, in detail, how to tap a hole on a turret lathe.

14-13. How are external threads generally cut on a turret lathe?

14-14. What is an automatic production lathe?

14-15. What is the difference between a vertical boring mill and a vertical turret lathe?

14-16. How are workpieces held on vertical lathes?

14-17. What type of workpieces are machined on a vertical lathe?

14-18. What types of operations are performed on vertical lathes?

CHAPTER **15**

Precision Hole Location — The Jig Borer

The problem of precision hole location is encountered by all craftsmen who are engaged in precision machine-tool work. Many holes in precision machine and engine parts, jigs, fixtures, and metal stamping dies must be machined to very close tolerances with respect to their diameters and their location on the workpiece. The hole-locating methods previously treated in this book satisfy the accuracy requirements for most ordinary holes; however, when the location of a hole must be very precise, other methods must be used.

Precision hole-location methods have been developed over the years by skilled craftsmen such as clockmakers, instrument makers, toolmakers, and machinists. Some of these methods are still used upon occasion and will be described later, in this chapter. An engineering solution to the precision hole-locating problem was the development of the jig borer which occurred at the end of World War I, independently, both in the United States and Switzerland. This machine tool is designed to machine holes that are precise, with respect to diameter and location. A bank of jig borers at work is shown in Fig. 15-1. Just prior to World War II, a further step was taken by The Moore Special Tool Company, when they developed the jig-grinding machine which is used to locate and grind holes in hardened metals to extremely close tolerances. More recently, coordinate measuring machines have been developed to accurately measure the location of holes.

Coordinate Dimensions—Transfer of Tolerances

The coordinate system of dimensioning is used on drawings for jig-borer work. This system of dimensioning is designed to be compatible with the coordinate axes of movement available on the jig borer. Other machines, particularly numerically controlled machines, are also designed to have table movement that is compatible with the coordinate axes of the coordinate dimensioning system. These dimensions are normally placed on the drawing by the draftsman; however, the machinist or the toolmaker is occasionally required to do this at the machine. It is worthwhile to learn the principles involved.

In the coordinate dimension system all dimensions are given from two reference axes that are perpendicular to each other. The position of these axes sometimes corresponds to the side of the workpiece, as at B, Fig.

324 PRECISION HOLE LOCATION—THE JIG BORER Ch. 15

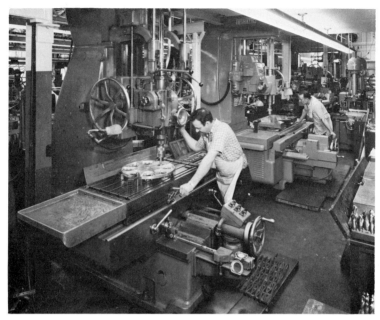

Courtesy of The Fosdick Machine Tool Company

Fig. 15-1. A bank of jig-boring machines in a tool-making department.

15-2; although this is not always the case. At A, Fig. 15-2, the dimensions are drawn in the conventional manner, while at B, they are drawn using the coordinate dimension system. Having the top edge and the left edge, or the upper left-hand corner, act as a reference surface, agrees with the positions of the scales on many jig borers, although there are some differences in this respect. The dimensions for the coordinate system for parts to be jig bored is usually given to .0001 inch. Tolerances are usually not specified, for it is understood that the jig-borer operator will locate each hole to the limit of accuracy of the machine. This is the accepted practice for tool and die work. The dimensions can be transferred from the conventional system of dimensioning to the coordinate system by simply adding or subtracting, as shown in Fig. 15-2.

All of the dimensions that must be accurately held on precision machine and engine parts are usually given a tolerance. When such dimensions must be changed from the conventional system of dimensioning to coordinate dimensions, the tolerances must also be included. The tolerances of the dimensions on the coordinate system of dimensioning will not be the same as on the conventional system of dimensioning from which they were derived. The new tolerances on the coordinate system of dimensioning must be carefully calculated, otherwise serious errors will result. In most instances, changing from the conventional to the coordinate system of dimensioning will reduce the tolerances.

Ch. 15 PRECISION HOLE LOCATION—THE JIG BORER

The principle of the transfer of tolerances can be stated as follows:

When transferring tolerances from the conventional dimensioning system to the coordinate dimensioning system, the sum of the tolerances of any pair of dimensions on the coordinate system must not exceed the tolerance of the dimension that they replace on the conventional system.

This principle can best be learned by following an example:

The $2.000^{\pm.005}$ dimension between two holes shown at A, in Fig. 15-3 is replaced by the $1.000^{\pm.0005}$ and the $3.000^{\pm.0045}$-inch dimensions in the coordinate systems shown at C, in Fig. 15-3. The sum of the tolerances in view C; i.e. ($\pm.0005$ and $\pm.0045$) must not exceed $\pm.005$-inch, otherwise the original tolerance of the $2.000^{\pm.005}$ dimension can be exceeded and the workpiece will not meet the original dimensional requirements. It is

Fig. 15-2. (Top) A. Conventional dimensions. (Bottom) B. Coordinate dimensions.

Fig. 15-3. (Top) A. Conventional dimensions—mixed tolerances. (Center) B. Conventional dimensions—all equal, bilateral tolerances. (Bottom) C. Coordinate dimensions.

important to realize that the sum of the new tolerances on the coordinate dimensions may be less than the original tolerance which they replace; however, they must never exceed the original tolerance.

It is best to follow a set procedure in order to transfer dimensions and tolerances from a conventional to a coordinate dimensioning system. This procedure is outlined below, and using the $2.125^{+.003}_{-.001}$ dimension in A, Fig. 15-3, as an example:

1. Where necessary, change the tolerances so that each dimension has an equal bilateral tolerance.

 In order to maintain the original dimensional requirement, the corresponding dimension must also be changed. The $^{+.003}_{-.001}$ tolerance of the 2.125 dimension is changed to ±.002, which is an equal bilateral tolerance. (Note that the total tolerance of .004 inch has not been changed.) The original dimensional requirements are maintained if the dimension is changed from 2.125 inches to 2.126 inches. The new dimension, $2.126^{\pm.002}$ is shown in view B. (Note that the dimension has not really changed, since in both cases, the limiting dimensions are 2.128 and 2.124 inches.)

2. Transfer the revised basic dimensions to the coordinate dimension system.

 The original basic dimension was 2.125 inches, which was revised in Step 1 to 2.126 inches. In the coordinate system this dimension is replaced by the .8750- and the 3.0010-inches dimensions. (The 3.0010 dimension is obtained by adding .8750 + 2.126 inches.)

3. Transfer the tolerances.

 The sum of the tolerances of the .8750 and the 3.0010 dimensions must not exceed the tolerance of the 2.126 dimension, or ±.002 inch, which is the available tolerance. If the available tolerance is to be distributed equally, then each new dimension will have ±.0010 tolerance. These dimensions will then read: $.8750^{\pm.0010}$ and $3.0010^{\pm.0010}$.

Sometimes the tolerance of a dimension on the coordinate-system drawing is affected by more than one dimensional requirement. In this case, the final tolerance used must be that which fulfills all of the requirements. For example, consider the $.502^{\pm.002}$ dimension in view B. It is replaced by the 3.0010 and the 3.5030 dimensions in view C. If the 3.0010 and 3.5030 dimensions are each given a ±.001 tolerance, the sum of their tolerance would not exceed the original ±.002 tolerance and, presumably, the requirements for the transfer of tolerances would be met. However, the 3.5030 dimension together with the 4.6280 dimension replaces the 1.125 dimension in view B, which has a tolerance of only ±.001 inch. Thus, the sum of the tolerances on the 3.5030 and the 4.6280 dimensions cannot exceed ±.001 inch which, when divided equally,

amounts to ±.0005 inch. The 3.5030-inch dimension must therefore be given the lesser tolerance of ±.005 inch.

For this reason, when transferring the tolerance it is usually best to start with the smallest tolerance. Also note that the sum of the tolerances replacing the .502$^{\pm.002}$ dimension is less than the ±.002 inch. This is satisfactory since the sum of the new tolerances does not *exceed* the original tolerances.

When a tolerance is bound by a small tolerance, as in the case of the .502 dimension, it is sometimes possible to increase an adjacent tolerance on the coordinate dimension drawing. For example, the 2.000-inch dimension in view B, which has a tolerance of ±.005 inch is replaced by a 1.0000- and a 3.0000-inch dimension on the coordinate drawing. Each of these dimensions could be given a tolerance of ±.0025 inch; however, the tolerance of the 1.0000-inch dimension on the coordinate drawing is bound by the requirement of the ±.001 tolerance of the 1.000 dimension shown at the right side, in view B. Thus, the 1.0000 dimension in view C, together with the 2.0000 dimension, must have a total dimension tolerance of ±.0010 or ±.0005 inch on the 1.0000-inch dimension. Since the sum of the tolerances of the 1.0000 and the 3.0000 dimensions (view C) can be ±.005 inch, the tolerance of the 3.000 dimension can be increased to ±.0045 inch.

The basis for the determination of the final tolerances for the coordinate dimensions (C, Fig. 15-3) is summarized below:

.8750$^{\pm.0010}$: The ±.0010 tolerance together with the ±.0010 tolerance of the 3.0010 dimension is required to maintain the ±.002 tolerance for the 2.126 dimension.

3.0010$^{\pm.0010}$: See requirements for the .8750$^{\pm.0010}$ dimension given above.

3.5030$^{\pm.0005}$: The ±.0005 tolerance together with the ±.0005 tolerance of the 4.6280 dimension is required to maintain the ±.001 tolerance for the 1.125$^{\pm.001}$ dimension.

4.6280$^{\pm.0005}$: See requirements for the 3.5030$^{\pm.0005}$ dimensions given above.

1.0000$^{\pm.0005}$: The ±.0005 tolerance together with the ±.0005 tolerance of the 2.000 dimension is required to maintain the ±.001 for the 1.000$^{\pm.001}$ dimension given at the right, in view B.

2.0000$^{\pm.0005}$: See requirements for the 1.0000$^{\pm.0005}$ dimension given above.

3.0000$^{\pm.0045}$: The .0045 tolerance, together with the ±.0005 tolerance of the 1.0000 dimension in view C, is required to maintain the ±.005 tolerance for the 2.000$^{\pm.005}$ dimension in view B.

Boring

Because of its importance in precision hole location, a review of the boring process is helpful at this time. In Fig. 15-4, the boring tool shown

Ch. 15 PRECISION HOLE LOCATION—THE JIG BORER 329

is rotating about the axis of the spindle of a jig-boring machine. The point of the cutting tool is always the same distance from the axis of rotation of the spindle; therefore, the cylindrical surface of the hole that is generated must be concentric with the axis of the spindle. For this reason, single-point boring is the most accurate method of finishing holes in metals that can be cut with single-point cutting tools.

Courtesy of The Moore Special Tool Company

Fig. 15-4. Single point boring—the most accurate hole-locating method. The surface is generated by rotation about the axis. Either the workpiece or tool may rotate.

The boring tool shown in Fig. 15-4 is held in an offset boring-head which can be very accurately adjusted, radially toward or away from the center of rotation, in order to bore holes of different diameters. With normal care it is not difficult to bore a hole to a tolerance in the order of "tenths," with a precision boring-head.

If the hole that was originally drilled is not concentric with the axis of the spindle, the boring tool will take a heavier cut on one side of the hole than on the other side. The unequal cutting force created by this cut will deflect the bar more on the side where the heavier cut is taken, thereby causing the bored hole to be eccentric with respect to the axis of the spindle. This effect can be eliminated by grinding the boring tool correctly, by using a sharp tool, and by taking light cuts when finishing the hole to size.

In the situation just described, the boring tool was rotating. The same end-result will also occur when the work is rotating, as in a lathe. The point of the cutting tool will always be at the same distance from the axis of rotation of the workpiece, which is determined by the axis of rotation of the lathe spindle. The cylindrical surface that is generated as the tool feeds through the bore must then be concentric with the axis of rotation of the spindle.

Precision Hole Location

The shapes and sizes of the workpiece in which precision holes are located are almost infinite. Rectangular-shape plates, such as shown in Fig. 15-5, are frequently encountered in tool and die work. Because of the availability and the ease with which this shape can be worked, it is used to construct components for stamping dies, jigs, fixtures, and gages. This shape, therefore, is used to illustrate and explain the methods of precision hole location. Four proven and practical methods of hole location will be explained in the following pages; in addition, two methods of locating holes in hardened workpieces will be discussed. It should be emphasized that the use of the jig borer, when available, is not only the preferred, but the superior method of precision hole location.

Precision Layout and Drilling Procedure

Prepare the workpiece. The workpiece is machined so that the two faces are parallel to each other and "square" with respect to all sides, or edges, as they are often called in the toolroom. Two of the edges are machined to be exactly square with each other; these two edges will serve as reference surfaces. Sometimes it is convenient to machine all four edges square to each other. The machining can be done on a milling machine, a shaper, or on a surface grinder. It is often more convenient to rough machine on a milling machine or on a shaper, and then to finish grind the part on a surface grinder. A metal-cutting band saw will sometimes be best for rough machining this part.

Make a precision layout. The face of the workpiece is painted with a blue layout dye and the lines are scribed with precision layout tools on the required locations. The precision layout tools are shown at B, Fig. 15-5. Shown are a vernier height-gage, precision gage-blocks, and a universal precision gage, or "planer" gage, as it is also called. The vernier height-gage has a scriber attached to the movable arm which can be set at any desired distance above the face of the surface plate. With care, layout lines can be scribed to an accuracy of .001 inch with the vernier height-gage. The slider of the planer gage can be accurately positioned anywhere along the incline. The top of the slider is perfectly parallel to its base and serves as a reference surface. It is positioned to the desired height above the surface plate by measuring across the top of the slider and the base with a micrometer caliper. A more accurate method is to transfer the height of a stack of precision gage-blocks to the top of the slider. This is done by holding a dial test indicator in a fixed vertical position on an

Ch. 15 PRECISION HOLE LOCATION—THE JIG BORER 331

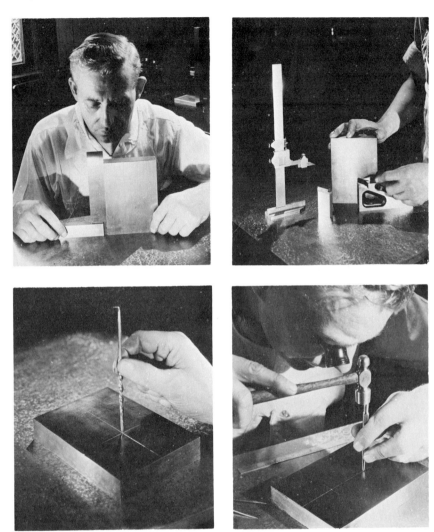

Courtesy of The Moore Special Tool Company

Fig. 15-5. (Upper left) A. Squaring-up work preparatory to laying out. The edges of the work will be used as a reference and must be accurately squared. (Upper right) B. Commonly employed layout tools. (Note wedge-shape scribers.) (Lower left) C. Locating the center of the hole with rotating, hand prick-punch with an acute-point angle. (Lower right) D. Carefully prick-punching the work with standard prick-punch and hammer.

indicator holder or on the movable arm of the height gage. The indicator is positioned over the gage blocks and set to read zero. It is then positioned over the planer gage which is adjusted until the indicator again reads zero. A wedge-shape scriber, which has had its bottom surface

Courtesy of The Moore Special Tool Company

Fig. 15-6. Error introduced by shape of scribed line. This error is reduced by keeping the scribed line shallow and by using a rotating, hand prick-punch with an acute-point angle prior to using a regular prick-punch.

lapped perfectly flat, is placed on the planer gage to scribe the layout lines. The wedge-shape scriber can also be placed directly on the stack of gage blocks to scribe the line if an accuracy to "tenths" is required and a planer gage is not available. This method, however, is hard on the gage blocks and should be used sparingly. Lines are thus scribed by whatever precision method is most suitable. The coordinate dimension layout for the holes is obtained by resting the two squared reference edges on the layout table. The scribed lines should be kept as shallow as possible.

Prick-punch marks are carefully made at the intersection of the layout lines where the holes are desired. When an ordinary prick punch is placed in a line that has been scribed by a wedge type scriber a small error will result, as shown in Fig. 15-6. In order to minimize this error, the prick-punch mark should be started with a hand prick-punch as shown at C, Fig. 15-5. A hand prick-punch has a more acute point than a regular prick-punch. It is made from a piece of square, high-carbon tool steel

that is twisted as shown, and hardened at the point. The hand prick-punch is carefully placed at the intersection of the layout lines, using the sense of touch to feel the intersection with the point of the scriber. It is possible to feel the scriber into the line. When the hand prick-punch is in position it is rotated by sliding the fingers down along the twisted portion of the scriber. After the hand scriber has been used, the prick-punch mark is enlarged with a standard prick-punch and hammer as shown at D, Fig. 15-5. This must be done carefully, using a magnifying glass to be sure that the original position is not disturbed. Also, the prick punch must be accurately ground to have a point in the center of the punch. The hammer blow should be light, making a prick-punch that is approximately $\frac{1}{64}$ inch in diameter. A circle may now be scribed with dividers although this is not always necessary, especially with smaller holes. The circle is not used to locate the hole. When a circle is scribed on precision layouts, the witness marks are usually omitted. On die and tool work they may mar a surface that must be finished smooth or they may even form the nucleus of a crack if the part is to be heat treated.

The layout is now completed. When done with skill and care, accurate results can be expected.

Drill and counterbore the hole from the layout. A large drill should never be started in a small prick-punch mark. The prick-punch mark will be quickly obliterated and the drill will drift into another location. Also, a center drill should not be used. The center drill is rigid and a slight misalignment between the center drill and the prick-punch mark will cause the drill to start the hole at the side of the prick-punch mark,

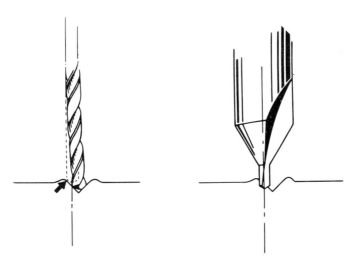

Courtesy of The Moore Special Tool Company

Fig. 15-7. Starting the hole with a small slender drill. Slender drill will bend and follow prick-punch mark. Center drill will cut into the side of prick-punch mark instead of following it.

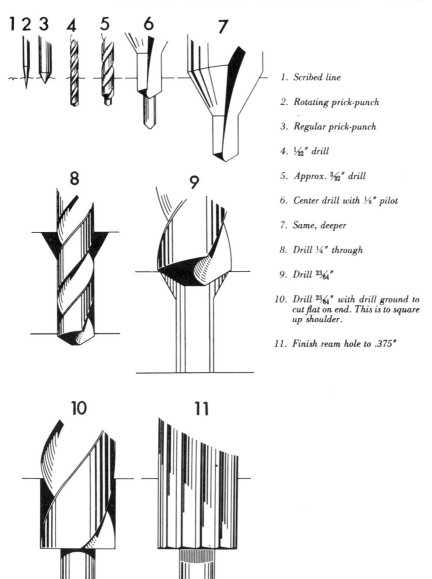

1. Scribed line
2. Rotating prick-punch
3. Regular prick-punch
4. 1/32" drill
5. Approx. 3/32" drill
6. Center drill with 1/8" pilot
7. Same, deeper
8. Drill 1/4" through
9. Drill 23/64"
10. Drill 23/64" with drill ground to cut flat on end. This is to square up shoulder.
11. Finish ream hole to .375"

Courtesy of The Moore Special Tool Company
Fig. 15-8. Procedure for drilling hole from layout.

as in Fig. 15-7, with a consequent error in the hole location. The best start is made when the prick punch entirely surrounds the drill. An ordinary twist drill, having a diameter of about $1/64$ or $1/32$ inch, should be used to start the hole. The position of the hole is determined by the prick-punch mark and any misalignment of the drill will be compensated for

Ch. 15 PRECISION HOLE LOCATION—THE JIG BORER 335

by the flexibility of the drill, as shown in Fig. 15-7. The slender twist drill will bend and center itself in the prick-punch mark, whereas the rigid center drill will not. Since the slender starting drill will bend, it will not drill a vertical hole, and it should not be permitted to go too deep. The entire sequence of starting the hole is outlined in Fig. 15-8.

The starting drill should be followed by another, slightly larger and stiffer drill, say 3/32 diameter, which is followed by a center drill having a 1/8-inch pilot. The center drill is drilled to the depth as shown at 7, Fig. 15-8. A 1/4-inch drill is then used to drill through the part. The hole is further enlarged with a 23/64-inch drill which is followed by a 23/64-inch drill that has been ground to cut flat on the end in order to square the shoulder of the counterbore. This operation is omitted if the hole is a through hole. The final operation is to finish ream the hole to size.

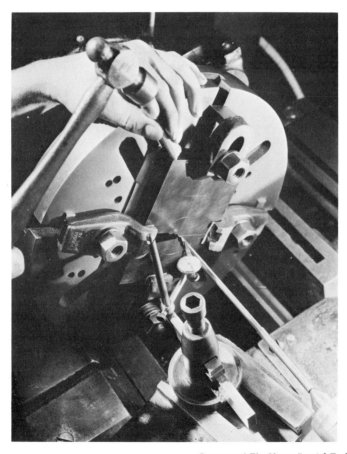

Courtesy of The Moore Special Tool Company

Fig. 15-9. Aligning the work with the axis of rotation of the lathe using a pump center and an indicator.

Precision Layout, Drilling, and Boring Procedure

1. Prepare the workpiece.
2. Make a precision layout. (Steps 1 and 2 are done as previously described and will, therefore, not be repeated.)
3. Mount the workpiece on a lathe faceplate. The workpiece is clamped to the faceplate, Fig. 15-9, in preparation for drilling and boring. On parts of this size it is usually most convenient to remove the faceplate to mount the work as shown in Fig. 15-29. A sheet of paper placed between the faceplate and the workpiece serves to reduce the tendency of the part to move, particularly when it is subjected to large cutting forces. Soft metal shims placed between the strap clamps and the work will prevent the clamps from marring the surface of the work.
4. Align the workpiece in the lathe. The point of a pump center, Fig. 15-10, is placed in the prick-punch mark corresponding to the hole to be drilled, as shown in Fig. 15-9. The other end is placed on the tailstock center of the lathe and the tailstock is adjusted so that the pump center is held in position with a light spring tension. The contact point of a dial test indicator is positioned against the pump center close to the point. When the lathe spindle is slowly turned by hand, the movement of the indicator hand will show whether or not the prick-punch mark is in the center of rotation. The work is tapped with a hammer and a soft metal bar, as shown, until the indicator hand does not move when the work is turned. The work is then tightly clamped to the faceplate.
5. Drill the hole. The prick-punch mark has served its purpose—that of aligning the part on the faceplate. It is no longer essential to the job. The hole should be started with a center drill. This should be followed by a ¼- or ⁵⁄₁₆-inch drill which is used to drill the hole through the workpiece. The hole is enlarged by drilling with successively larger drills until it reaches the size for boring. Too much metal should not be removed with any one drill in order to keep the cutting forces from being large enough to move the piece on the faceplate. Sufficient stock should be left for boring to allow several boring cuts to be taken. This can vary from ⅛ to ¼ inch or more.
6. Bore the hole to size. Figure 15-11 illustrates the holes being bored to size. The trial cut procedure is used to bore the hole to size. A trial cut is taken in the bore and the diameter is carefully measured. The boring tool is then advanced into the workpiece the correct amount, and another trial cut is taken. If the measurement

Fig. 15-10. Pump center.

Ch. 15 PRECISION HOLE LOCATION—THE JIG BORER

Courtesy of The Moore Special Tool Company

Fig. 15-11. Boring the hole in the lathe with the work mounted on a faceplate.

of the second trial cut shows the diameter to be correct, the cut is completed; if not, a further adjustment is made before completing the cut. The final finishing cut should not be started until the trial cut shows that the diameter being cut is correct.

Several light boring cuts should be taken through the workpiece to make certain that the hole will be concentric with the axis of rotation of the lathe. When using this procedure the hole should be bored to size; it should not be finished by reaming because inaccuracy may result due to any slight misalignment between the spindle axes of the headstock and the tailstock.

Toolmakers Buttons, or the Buttoning Procedure

Toolmakers buttons, Fig. 15-12, are hardened, ground, and lapped collars, or buttons, with the ends ground perfectly square with the cylindrical surface. Thus, when the buttons are clamped in place with the

Courtesy of The Moore Special Tool Company

Fig. 15-12. Toolmakers buttons. The cylindrical surface is hardened, ground, and lapped. The end faces must be exactly perpendicular to the cylindrical surface.

$\frac{1}{8}$-40 machine screw, the cylindrical surface will be perpendicular to the workpiece. The oustide diameter of the buttons is made in several sizes, namely, in .300, .400, and .500 inch. It is not good practice to mix the diameters of the buttons in any one setup as this can result in errors. The inside diameter of the buttons is slightly larger than the screw in order that the button can be moved around the screw.

The buttoning procedure is a good procedure that is frequently used when a jig borer is not available. This procedure is generally more reliable than those previously described. A number of buttons can be attached to the workpiece, simultaneously, in order to locate holes in the workpiece. For simplicity, the description to follow will concern itself with only one hole:

1. Prepare the workpiece. It is machined square and parallel all over, using methods previously mentioned.
2. Lay out the part. When buttoning, a precision layout is not required although the precision layout methods described are often convenient to use.
3. Drill and tap the button screw holes. Each hole is drilled with a No. 38 drill, using ordinary care. All holes are then tapped with a $\frac{1}{8}$-40 tap.
4. Attach the buttons to the workpiece. The buttons should be clamped lightly but firmly so that they can be moved with a light tap of a hammer.
5. Locate each button on the workpiece. The part is placed on the surface with one of the reference edges down. (See Fig. 15-13.) Attach a dial test indicator to the vernier height-gage, as shown. With the sliding arm of the height gage adjusted so that the vernier scale reads zero, position the contact points of the indicator against the surface plate and adjust the indicator to also read

Ch. 15 PRECISION HOLE LOCATION—THE JIG BORER 337

Courtesy of The Moore Special Tool Company

Fig. 15-11. Boring the hole in the lathe with the work mounted on a faceplate.

of the second trial cut shows the diameter to be correct, the cut is completed; if not, a further adjustment is made before completing the cut. The final finishing cut should not be started until the trial cut shows that the diameter being cut is correct.

Several light boring cuts should be taken through the workpiece to make certain that the hole will be concentric with the axis of rotation of the lathe. When using this procedure the hole should be bored to size; it should not be finished by reaming because inaccuracy may result due to any slight misalignment between the spindle axes of the headstock and the tailstock.

Toolmakers Buttons, or the Buttoning Procedure

Toolmakers buttons, Fig. 15-12, are hardened, ground, and lapped collars, or buttons, with the ends ground perfectly square with the cylindrical surface. Thus, when the buttons are clamped in place with the

Courtesy of The Moore Special Tool Company

Fig. 15-12. Toolmakers buttons. The cylindrical surface is hardened, ground, and lapped. The end faces must be exactly perpendicular to the cylindrical surface.

$\frac{1}{8}$-40 machine screw, the cylindrical surface will be perpendicular to the workpiece. The oustide diameter of the buttons is made in several sizes, namely, in .300, .400, and .500 inch. It is not good practice to mix the diameters of the buttons in any one setup as this can result in errors. The inside diameter of the buttons is slightly larger than the screw in order that the button can be moved around the screw.

The buttoning procedure is a good procedure that is frequently used when a jig borer is not available. This procedure is generally more reliable than those previously described. A number of buttons can be attached to the workpiece, simultaneously, in order to locate holes in the workpiece. For simplicity, the description to follow will concern itself with only one hole:

1. Prepare the workpiece. It is machined square and parallel all over, using methods previously mentioned.
2. Lay out the part. When buttoning, a precision layout is not required although the precision layout methods described are often convenient to use.
3. Drill and tap the button screw holes. Each hole is drilled with a No. 38 drill, using ordinary care. All holes are then tapped with a $\frac{1}{8}$-40 tap.
4. Attach the buttons to the workpiece. The buttons should be clamped lightly but firmly so that they can be moved with a light tap of a hammer.
5. Locate each button on the workpiece. The part is placed on the surface with one of the reference edges down. (See Fig. 15-13.) Attach a dial test indicator to the vernier height-gage, as shown. With the sliding arm of the height gage adjusted so that the vernier scale reads zero, position the contact points of the indicator against the surface plate and adjust the indicator to also read

zero. The top of the surface plate is now the zero reference plane for the height gage and the workpiece. The height gage is set to read the distance from the reference edge to the center of the hole plus one-half of the diameter of the toolmakers button. The indicator contact point is now run over the top of the button and the maximum reading is noted. The button is positioned by tapping it lightly with a hammer and a piece of soft metal, as shown in Fig. 15-13, until the maximum reading of the dial indicator is zero when it is passed over the button. The button is now located with respect to one reference edge and the procedure is repeated with respect to the other reference edge. The toolmakers button is located when the indicator reads zero over the button from each reference edge with the height gage correctly set for each position. It should now be tightened securely in position.

An alternate method of locating the buttons is to use a stack of precision gage blocks. The length of the stack should be equal to the distance over the toolmakers button as explained in the last paragraph. A dial test indicator is held by a height gage and zeroed *over* the precision gage blocks which are resting on the surface plate. The indicator is run over the top of the toolmakers button and the button tapped into position until the maximum indicator reading is zero, as before. Again, this must be done from

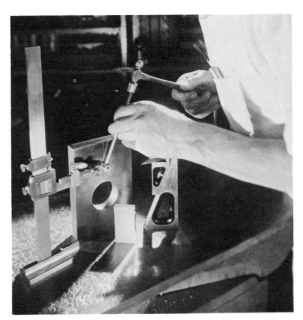

Courtesy of The Moore Special Tool Company

Fig. 15-13. Locating the button on the workpiece. This must be done in two directions from two reference edges.

Courtesy of The Moore Special Tool Company

Fig. 15-14. Aligning the workpiece in the lathe with a toolmakers button.

the two reference surfaces. The use of the precision gage-blocks makes it possible to locate the buttons to within "tenths" when this procedure is carried out with skill and care.

6. Mount the work on the faceplate. This is done as previously described. A counterweight, shown in Fig. 15-14, is used to balance the part and prevent unwanted vibrations from occurring.
7. Align the work on the faceplate. This operation is illustrated in Fig. 15-14. The dial test indicator is placed in the lathe as shown and its contact point is positioned against the toolmakers button. The faceplate is slowly turned by hand and the movement of the indicator hand is observed. The workpiece is adjusted by tapping it with a hammer and a piece of soft metal until the indicator hand does not move when the faceplate is turned at least one complete turn.
8. Remove the toolmakers buttons.
9. Center drill the hole. Use a large center drill that is consistent, of course, with the final size of the hole. The large center drill, being stiff, will tend to correct any eccentricity in the tapped holes used to clamp the toolmakers buttons.
10. Drill the hole. The drilling operation is done as previously described for the faceplate-mounted workpiece.
11. Bore the hole to size. This is also done as previously described.

Toolmakers buttons are also used to perform precision hole location operations on a milling machine. The buttons are located and clamped on the workpiece as has been described for lathe work. The workpiece is aligned and clamped on the table of the milling machine so that the holes to be produced can be reached by tools held in the spindle of the machine. A dial test indicator is attached to the milling machine spindle. Each hole is located by placing the contact point of the indicator against the toolmakers button and slowly rotating the spindle by hand. When the button is located on the axis of the milling-machine spindle, as shown by the indicator reading, it is removed. The hole is then machined by drilling with a twist drill, and boring with a boring tool. For convenience, the boring tool should be held in an offset boring head, Fig. 15-26, which provides an accurate radial adjustment of the boring tool in order to bore the hole to size.

The Jig Borer

The jig borer is designed specifically for doing precision hole-location work. Its design is a combination of the application of sound engineering principles and a thorough knowledge of toolmaking practices. The highest order of craftsmanship is required in its construction.

A precision jig-boring machine is shown in Fig. 15-15. The bed of the jig borer rests on the floor and the column is bolted to a pad machined on the top of the bed. The column supports the driving motor, the spindle drive mechanism, the power-feed mechanism, the quill which contains the spindle, and the spindle housing. Highly accurate, machine-tool-grade anti-friction bearings are used to mount the spindle in the quill. These bearings are pre-loaded in order to reduce the tendency for chatter to occur when boring. Care is taken in the construction of the machine to make certain that the up and down movement of the spindle is perpendicular to the table.

The table is mounted on the top of a saddle and the saddle is mounted on the bed. The saddle moves toward and away from the column on a machined and hand-scraped slide. It carries the table with it in this motion which is called the *transverse* table movement. A machined and hand-scraped slide on top of the saddle allows the table to move in a direction perpendicular to the transverse direction and this movement is called the *longitudinal* table movement.

The coordinate measuring system of a jig borer is designed to provide longitudinal and transverse table movements that are accurate to one-ten-thousandths (.0001) inch. On the Moore jig borer, shown in Fig. 15-15, the coordinate measuring system is based on the two lead screws (see Fig. 15-16) that actuate the table movements. These screws are by no means ordinary screws. They are hardened and ground to an accuracy of .0002 inch in 16 inches. A micrometer dial attached to the end of each lead screw is graduated to read one-thousandth (.001) inch. One ten-thousandth (.0001)-inch readings are obtained by a vernier scale which is placed adjacent to the micrometer dial. Two steel rules are attached

342 PRECISION HOLE LOCATION—THE JIG BORER Ch. 15

Courtesy of The Moore Special Tool Company

Fig. 15-15. Model No. 2, Moore Precision Jig-Borer.

to the machine and are placed parallel to each of the two table movements. These rules are used to obtain the coarse table settings.

Another coordinate measuring system which is used on many jig borers is shown in Fig. 15-17. This system is based on precision end measuring rods and a micrometer head which are placed in a trough that is parallel to the table movement. There is a trough for each table movement and, therefore, there are also two sets of precision end measuring rods and micrometer heads. The precision end-measuring rods are made in different lengths with one-inch increments. The subdivisions of an inch are obtained by the micrometer head which has a vernier scale that enables it to be set to an accuracy of one-ten-thousandth (.0001) inch. An adjustable stop is placed at one end of each trough. The function of this stop is to move the stack, consisting of the precision end-measuring rods and the micrometer head, along in the trough until it touches the contact point of a dial test indicator which is in a fixed position at the other end of the

Ch. 15 PRECISION HOLE LOCATION—THE JIG BORER 343

Courtesy of The Moore Special Tool Company

Fig. 15-16. System of measuring the coordinate movements of the jig borer table using precision lead screws.

Courtesy of The Fosdick Machine Tool Company

Fig. 15-17. System of measuring the coordinate movements of the jig-borer table using precision end-measuring rods and micrometer heads. Arrows show dial test indicators.

trough. The only time that the lengthwise adjustment of the adjustable stop is used is after the initial alignment of the reference axes of the workpiece with respect to the axis of the spindle. At this time, only the micrometer head is placed in the trough and the adjustable stop is used to move it along in the trough until the dial test indicator reads zero. At other times, the adjustable stop moves along with the component to which it is attached—either the table or the saddle—and when the table settings are made it pushes the stack until the dial test indicator hand reads zero. The dial test indicator has one-ten-thousandths-inch graduations; however, these graduations should not be used to obtain one-tenthousandth-inch table settings which are obtained by the setting of the micrometer head. The function of the dial test indicator is to place a uniform end-pressure on the end-measuring rods and the micrometer head, in order to hold them together and to provide a precise reference position for this stack in the trough.

Cleanliness and care must be observed when the precision end-measuring rods and the micrometer head are placed in the trough. The ends, where they contact each other, must be wiped clean. When a setting has been made, a check for cleanliness is made by slightly rotating each precision end-measuring rod and micrometer head in the trough, while observing the hand of the dial test indicator. If the hand moves, dirt is present. This dirt must be removed and the table setting made over again.

Several other coordinate measuring systems are used on different jig borers. It is not possible to describe all of these, in detail, within the scope of this book. The purpose of all these systems, however, is alike. It is to provide a method of aligning the spindle at the precise location on the workpiece where the hole is to be produced. The primary machining operations for which the jig borer is designed are drilling and boring. Operations such as reaming, counterboring, countersinking, and spot facing are also done on the jig borer. Light end-milling operations are sometimes performed on this machine although heavier cuts should be avoided. An air-driven, internal grinding attachment can be obtained for jig grinding. In addition to the production of holes in precise locations, the jig borer can be used to machine certain contours on dies and other tools.

Jig-Boring Practice

There are four basic steps which should be followed in sequence in order to locate and machine a hole on a jig borer. These steps are: 1. align and clamp the workpiece on the jig-borer table; 2. locate the two reference axes of the workpiece with respect to the jig-borer spindle; 3. locate the hole to be machined; and 4. drill and bore the hole to size. These steps will now be treated in detail.

Align and Clamp the Workpiece. The first consideration in placing the workpiece on the jig-borer table should be the relation of the coordinate measuring system of the jig borer to the coordinate dimensions on the drawing. Although the jig-borer table can be accurately positioned when

it is moved in any direction, the coordinate measuring system is designed so that the reading of the measurement is direct when the table is moved toward the left and toward the column. As shown in Fig. 15-18, moving the table forward toward the left has the same effect as moving the spindle toward the right. Likewise, moving the table toward the column has the same effect as moving the spindle away from the column, or down, when seen in the plan view. For this reason, the coordinate dimensions on drawings made for jig-borer work are made to read from left to right and from the top of the view toward the bottom. The reference axes, from which the coordinate dimensions are taken are, therefore, placed at the top and to the left in the view on the drawing which shows these dimensions.

If the workpiece is placed on the jig-borer table so that one reference axis is toward the column and the other reference axis is toward the left, the coordinate dimensions on the workpiece will read in the same direction as the effective spindle movement when the coordinate measuring system on the jig borer reads direct. In other words, when the workpiece is placed in this position, the coordinate setting of the jig-borer table can be made to correspond exactly to the coordinate dimensions on the drawing in order to locate a hole. This avoids any error that can occur when calculations are required in order to make the table settings.

Frequently, the position of the reference axes corresponds with two of the edges on the workpiece. These edges should be machined perpendicular to each other before the workpiece is placed on the jig borer. They can then be used to align the workpiece on the jig-borer table and to locate it with respect to the spindle. Such workpieces should be placed on the table with the reference edges positioned to the left and toward the column of the jig borer.

If the holes to be bored pass through the bottom of the workpiece it must be placed on precision parallel bars (see Fig. 15-19) in order to provide clearance for the cutting tools and to prevent the table from being damaged. The method of clamping the workpiece will, to a large extent, depend upon the shape of the workpiece. If parallels are used they should be placed directly under the clamps to prevent the force exerted by the clamps from bending the workpiece. If this occurs, the workpiece will spring back when the clamps are released with the possible result that the work done on the jig borer will be inaccurate.

Before the workpiece is firmly clamped in place it must be accurately aligned with respect to the longitudinal and transverse table movements. The most effective method of aligning the workpiece is to indicate along an edge of the workpiece that is parallel to or, preferably, that corresponds to a coordinate dimension reference line. As an example, the workpiece shown in Fig. 15-19 is a die block on which the two reference edges have been machined perpendicular to each other. The reference edges are placed to the left and toward the column of the jig borer. The die block is shown being aligned by indicating along the reference edge. The dial test indicator is held in the spindle of the jig borer and the table is trav-

Fig. 15-18. A. Position of table relative to spindle before table movement. B. Position of table after moving in conventional direction by direct reading of coordinate measuring system. Note that the "effective" movement of the jig-borer spindle is opposite in direction from the actual table movement.

Ch. 15 PRECISION HOLE LOCATION—THE JIG BORER 347

Courtesy of The Moore Special Tool Company

Fig. 15-19. Indicating the edge of a die block to align it with the longitudinal table movement.

ersed in the longitudinal direction. As the table is traversed the indicator reading is observed. The workpiece is adjusted until the indicator reading is the same for all positions.

Locate the Two Reference Axes of the Workpiece with Respect to the Spindle. The workpiece must now be positioned so that the reference axes of the coordinate dimensions are a known distance from the axis of the jig-borer spindle. On some workpieces the two reference axes intersect at the center of a previously machined hole. In this case, the work and the spindle can easily be aligned by placing an indicator in the spindle and indicating the hole. This will place the axis of the spindle on both of the reference lines on the workpiece. With this accomplished, the coordinate measuring system of the jig borer is adjusted to read zero, or some predesignated reading, from which the table settings to follow are made.

The die block in Fig. 15-19 is an example of a workpiece on which the reference axes correspond to two of the edges. These edges are located to the left and toward the column. They must now be positioned by moving

the table until they are at some known distance from the axis of the spindle. Each edge is "picked up" by one of the methods to be described further on. As one of the edges is picked up, the measuring system on the jig borer that is perpendicular to that edge is set to read zero or to a pre-designated reading. Then the other measuring system is zeroed, or set, at a predesignated reading after the other edge has been picked up. With this accomplished, the relationship between the workpiece and the spindle is established. The table settings for boring each hole can then be made to correspond with the coordinate dimensions on the drawing.

The procedure for picking up an edge should be thoroughly understood. Several procedures can be used. Four of the most common procedures will be described. The first method to be described requires no special tools, only a dial test indicator and a small precision parallel or gage block. This procedure is illustrated in Fig. 15-20. The dial test indicator is mounted on the spindle of the jig borer with its contact point slightly beyond the centerline of the spindle (to the right of the centerline when the indicator is in the position shown in the view at the left, in Fig. 15-20). The spindle is lowered and the table is moved until the contact point of the indicator touches the side or edge of the workpiece. The spindle is rotated back and forth a small amount, by hand, and the maximum indicator reading is noted. Then the spindle is raised and a small precision parallel or gage block is held against the workpiece. The contact point of the indicator is placed against the parallel or gage block, and the

Courtesy of The Moore Special Tool Company

Fig. 15-20. Procedure for picking up the edge of a workpiece using a dial test indicator and a gage block.

Ch. 15 PRECISION HOLE LOCATION—THE JIG BORER

Courtesy of The Moore Special Tool Company

Fig. 15-21. Procedure for picking up the edge of a workpiece using a Moore Edge Finder.

spindle is rotated as before. Again the maximum indicator reading is observed and this reading is compared to the previous reading. The table is adjusted until both maximum indicator deadings are the same. When this occurs the axis of the spindle will fall on the edge and the measuring system on the jig borer perpendicular to this edge is set to read zero.

A better method is to use the Moore Edge Finder which has been developed by The Moore Special Tool Company. It is shown in Fig. 15-21. This edge finder is a highly accurate angle block with a vertical locating surface which is below and in the center of two indicating lips that form opposite sides of a rectangular opening. The horizontal distance between these lips is .400 inch and between each lip and the locating surface it is .200 inch. A dial test indicator is mounted in the spindle of the jig borer. The contact point is brought against the two lips by rotating the spindle with one hand while the other hand is used to hold the locating surface of the edge finder against the edge of the workpiece. The maximum indicator reading at each lip is observed and the table is adjusted until the maximum indicator readings are equal. When this occurs, the axis of the spindle will fall on the edge and the jig-borer measuring system is set to zero as before.

The centering microscope, Fig. 15-22, is designed to be used as an edge finder. The eyepiece can be held stationary as the spindle is slowly turned. The magnification is such that "tenths" (.0001 inch) can be seen and the field of vision is large enough to show a fairly large area of the workpiece. Two crossed hairlines act as the center of the optical system. The center of the optical system should be checked, but when verified, it is only necessary to move the table until the edge of the workpiece coincides with one of the hairlines on the microscope in order to align the axis of the spindle on the edge of the workpiece.

A very popular edge finder is shown in Fig. 15-23. It can be obtained with a contact at one or at both ends which is held to the body by springs.

Courtesy of The Fosdick Machine Tool Company

Fig. 15-22. Centering microscope used to pick up the edge of a workpiece.

Courtesy of The L. S. Starrett Company

Fig. 15-23. (Left) Application of cylindrical contact of edge finder to pick up an edge. (Right) Application of pointed contact to pick up a scribed line.

Ch. 15 PRECISION HOLE LOCATION—THE JIG BORER

Only a very light pressure is required to slide the contacts along the end of the edge finder. The diameter of the cylindrical contact is .200 inch and this contact is always used to pick up the edge of a workpiece. The body of the edge finder is held in the spindle by a drill chuck or a collet chuck which must be rotating when the edge finder is used. The contact is intentionally offset so that it has a distinct wobble as it rotates. The table of the machine is then moved to bring the edge of the workpiece against the contact. As the edge of the workpiece continues to move toward the edge finder, the wobble of the contact will continually decrease until it ceases altogether. When this occurs, the table movement is stopped and the edge of the workpiece will be located at a distance equal to one-half of the diameter of the contact, or .100 inch, from the axis of the spindle. When this edge finder is used on a workpiece such as the die block, the measuring system on the jig borer perpendicular to this edge must be set to read .100 inch. From this reference position the table can be moved to locate a hole on the workpiece by setting it to the coordinate dimension given on the drawing.

The cone-point shape contact is used to pick up lines scribed on the workpiece as shown in Fig. 15-23. In this case the cone point must be made to rotate without wobble. This can easily be accomplished by touching the rotating cone point lightly, with an object such as a pencil. The alignment is then determined visually by lowering the cone point over the line.

Locate the Hole. The design of the coordinate measuring system on the jig borer will determine the procedure used to locate the workpiece so that the axis of the spindle will coincide with the axis of the hole to be machined. The table is moved an exact measured distance to the left and toward the column which is determined by the coordinate dimensions. On the Fosdick Jig Borer in Fig. 15-17, the total length of the precision end-measuring rods plus the micrometer-head setting in each trough should be equal to the coordinate dimension of the hole being bored. On the Moore Jig Borer, Fig. 15-16, the micrometer dial on each feed screw is set to zero when the two edges are picked up; however, the scales on the side are not necessarily set to zero. Usually, they are set at some full inch mark and this distance must be added to the coordinate dimension when positioning the table to locate the hole.

Drill and Bore the Hole. The sequence of operations used to produce a hole on the jig borer is based upon two basic principles: 1. A short, stiff drill, such as a center drill, that will not deflect when cutting should be used to start a hole whenever the work and the axis of the machine tool spindle are located at the exact position where the hole is wanted; 2. A single-point boring tool that is rotating about the axis of the machine tool spindle will generate or cut a surface that is concentric to the axis of rotation. Of course, a drill is required to open the hole so that the boring tool can enter and enlarge it.

The actual steps required to machine the hole are now listed in a numbered sequence.

352 PRECISION HOLE LOCATION—THE JIG BORER Ch. 15

Courtesy of The Moore Special Tool Company

Fig. 15-24. Spot-drilling holes in a die block on a jig borer.

1. Spot drill the hole with a center drill or a spotting drill. This operation is illustrated in Fig. 15-24. A conventional drill and countersink, or center drill, is recommended for this operation except when the holes to be machined are very small. In this case a special spotting drill, shown in Fig. 15-25, should be used. The spotting drills must be kept sharp because they are not intended

Courtesy of The Moore Special Tool Company
Courtesy of The Dumont Drill Company

Fig. 15-25. Spot drilling for producing very small holes using a spotting drill. (Right) Special spotting drill.

to penetrate deeply into the workpiece. Sometimes all of the holes to be produced in a part are spot drilled before they are drilled and bored. Some jig borer operators will spot all of the holes lightly and then re-spot them to depth in order to insure against making errors.

2. Drill the hole. The initial hole should be drilled with a twist drill that will follow the countersink made by the center drill. Although larger drills are sometimes used, it is best not to start with a drill in excess of about one-half inch in diameter. If required, the hole can then be enlarged by drilling with successively larger drills. The drill must leave enough metal in the hole to allow the boring tool to finish the hole to size.

Heat will be generated by the drilling operation, which will cause the work to expand, especially when much metal is removed by larger size drills. If the finish-boring operations are conducted while the work is very hot, the contraction occurring when the workpiece cools can cause inaccuracy. For this reason it is sometimes good practice to drill all of the holes first, allowing the workpiece to cool before the holes are bored to size. A procedure that is used at times is to spot drill all of the holes on the jig borer, rough drill all of the holes on a conventional drill press, and then finish the holes in a jig borer by boring.

Holes less than one-sixteenth inch in diameter are very difficult or even impossible to bore with a single-point tool. The first step in producing these small holes is to drill a spot with a spotting drill. The diameter of this spot should be somewhat less than the diameter of the hole. One method of finishing the hole is to drill it to size with a twist drill. A small drill will follow the spot made by the spotting tool quite accurately. Another method of finishing the hole after it has been spotted, is to drill it first with a twist drill that is one size smaller than the finish size of the hole. The hole is then finished with a twist drill of the size of the finished hole. This last drill is very carefully honed at the corners where the lip and the margin meet. Only a small amount of honing should be done on the two corners and care must be exercised to hone a relief angle so that the drill will cut properly. Small holes may be finished by reaming after they have been spotted and drilled. This is frequently done when no suitable drill is available corresponding to the diameter of the hole. In such cases it is not uncommon to make a special reaming tool or to grind a standard reamer to the desired diameter.

3. Bore the hole to size. The offset boring head in Fig. 15-26 provides an accurate and reliable method of adjusting the boring tool to cut the desired diameter on the workpiece. There are several designs of offset boring-heads but they all serve to radially offset the boring tool. The amount of offset can be accurately controlled

Courtesy of The Moore Special Tool Company
Fig. 15-26. Offset boring head.

by graduations which are usually in one-ten-thousandths (0.0001) inch. On some offset boring-heads the boring tool can be fed radially outward, continuously, which is done when taking external facing cuts around a hole, when facing internal shoulders, and when cutting grooves inside the hole.

A boring operation is illustrated in Fig. 15-27. The trial-cut procedure is used to perform the boring operation. A short trial cut is taken inside the bore and the resulting diameter is measured. Using the graduations on the offset boring-head, the boring tool is adjusted to the distance required to bore the hole to the desired

Ch. 15 PRECISION HOLE LOCATION—THE JIG BORER

size. If the hole is close to the finish size, another trial cut is taken and a measurement is made to verify the previous setting of the offset boring-head. The hole is then bored to the required depth.

When the diameter of the hole must be greatly enlarged by the boring tool, one or more rough boring cuts are taken to remove the bulk of the metal. If the workpiece becomes very hot in this process it should be allowed to cool before the finish boring cuts are taken. Before the hole is bored to the finish size, one or two shallow cuts should be taken through the hole with a single-point boring tool. These cuts are important because they remove any error in the location of the hole brought about by the operations performed beforehand. After a sufficient number of light boring cuts have been taken through the hole, it is finished to size with the single-point boring tool.

Sometimes, a reamer is used to finish the hole to size; however, before this tool is used, the location of the hole must be established by taking several light boring cuts with a single-point boring tool, as before. The hole should be bored .008 to .030 inch undersize and the reamer is used to remove the excess metal in the hole to bring it to the required size. While this procedure is faster, it is not quite as accurate as when the hole is finished by boring.

Courtesy of The Moore Special Tool Company

Fig. 15-27. Finish boring the hole to size on the jig borer.

Jig Grinding and the Jig Grinder

Hardened metals cannot be cut with an ordinary single-point cutting tool. The most effective way of machining these metals is by grinding. Tools and dies must frequently be finish-ground after they have been hardened. Certain precision machine parts must also be ground after they are hardened. Holes in hardened machine parts are frequently ground on internal grinding machines, or on cylindrical grinding machines, using an internal grinding attachment. Although holes in hardened tools and dies are, on occasion, also ground on these machines, the size and nature of these components does not usually lend them to this method of grinding. Holes in hardened tools and dies can also be ground on an engine lathe (Fig. 15-31) but a preferable method is to grind them on a jig-grinding machine (Fig. 15-32) if one is available.

A hole must exist before it can be enlarged by grinding; thus, the holes that are to be ground have been drilled and bored into the workpiece before it is hardened. The heat treatment used to harden the workpiece usually causes a small change in the size of the hole and possibly, in its roundness. The grinding operation is used to correct these errors and to reestablish the precise location of the hole.

Enough stock must be left in the hole by the drilling and boring operation to allow the hole to be finished to size by grinding. The amount of grinding stock to be left in the hole will depend upon the size of the hole, the depth of the layer of hardened metal surrounding the hole, and the shape of the workpiece. Generally, but not invariably, the amount of stock left for grinding should be .008 to .015 inch for small holes and .016 to .030 inch for larger holes. Hardened tools and dies, such as the blanking die in Fig. 15-30, should be ground all over on a surface grinder before the holes are ground. These surfaces should be ground square and parallel to each other.

Usually, the existing holes in tools and dies form the basis from which the final hole location is determined. Thus, the first step is to measure the location of the existing holes. Errors caused by the heat-treating operation or by previous machining operations will thus be detected and small errors can be corrected by the hole-grinding operation. Sometimes this is done by slightly shifting the location of each hole provided that there is sufficient grinding stock in each hole.

One method of locating the holes for grinding is by buttoning. The toolmakers buttons are fastened over the existing holes by one of the methods shown in Fig. 15-28. The buttons for all of the holes can be located from the two reference edges on the side of the workpiece by using the procedure described earlier in this chapter. In other instances, one of the holes is first ground to size and the remaining holes are then located from this hole.

A general outline of the procedure used is given using a blanking die as an example. The first step is to mount this die on a faceplate which has been removed from the lathe. (See Fig. 15-29.) It is easier to mount the

Ch. 15 PRECISION HOLE LOCATION—THE JIG BORER 357

Courtesy of The Moore Special Tool Company

Fig. 15-28. Methods of fastening toolmakers buttons over existing holes.

die when the surface of the faceplate is in a horizontal position. After the faceplate is attached to the lathe spindle, the blanking die is accurately located by indicating the hole as shown in Fig. 15-30. The hole is then ground to size using a tool-post grinding attachment. This operation, illustrated in Fig. 15-31, also serves to locate this hole in the die. A

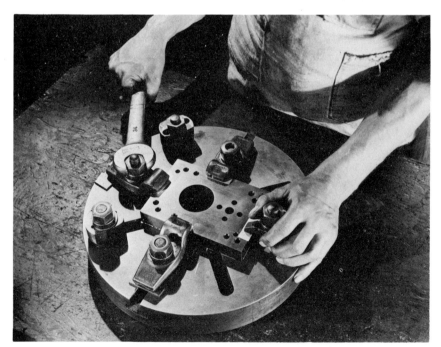

Courtesy of The Moore Special Tool Company

Fig. 15-29. Clamping the workpiece to a lathe faceplate. The faceplate is removed from the lathe spindle to make this job easier.

Courtesy of The Moore Special Tool Company

Fig. 15-30. Locating a blanking die on a faceplate by indicating the hole.

Courtesy of The Moore Special Tool Company

Fig. 15-31. Locating a hole in a hardened die by mounting it on a faceplate and grinding it in a lathe using a toolpost grinding attachment.

slight taper, or "draft" as it is called, of about one-quarter degree is then ground in the bottom of the hole to provide a clearance so that, when the die is used, the stock can fall through the die. It is ground by traversing the grinding wheel through the hole with the compound rest which is positioned at the draft angle.

The blanking die is then removed from the faceplate and toolmakers buttons are attached roughly in place over the remaining holes. It is placed on a precision surface-plate and each button is located by measuring from the hole that has been ground. The blanking die must be mounted on the faceplate again and located by indicating one of the toolmakers buttons as shown in Fig. 15-14. After the button is located it is removed. The hole is ground to finish size and the draft is ground in the back of the hole. This procedure is repeated until each hole in the die has been ground.

An increase in accuracy and a savings in time are possible if the job described above is performed on a jig grinder. This machine, illustrated in Fig. 15-32, was developed by The Moore Special Tool Company as an answer to the jig-grinding problem. It is very similar in construction to the jig borer. The longitudinal and transverse table positions are obtained with the aid of a precise coordinate measuring system. On the Moore jig grinder, in Fig. 15-32, this system is based on two precision lead screws. For each direction of table movement the coarse table position is obtained by reading a steel rule, and the precise setting is obtained by reading a micrometer dial at the end of the screw. A vernier scale adjacent to the micrometer dial makes it possible to read one-ten-thousandth (.0001) inch as in the case of the jig borer. Other jig grinders have different systems of measuring the coordinate table setting. One system that is frequently used employs precision end-measuring rods and micrometer heads in the same manner as is used on jig borers.

The axis of rotation of the main spindle of the jig grinder determines the center of the hole that is ground. The main spindle rotates but it does not move up and down except when the position of the entire spindle housing is adjusted. The continuous, up and down traverse motion of the grinding wheel, which is required in grinding a hole, is provided by a vertical slide which is located inside the main spindle. An eccentric cross slide, which cannot be seen in Fig. 15-32, is attached to the vertical slide. The eccentric cross slide provides a means of adjusting the grinding wheel radially outward in order to grind the diameter of the hole to the desired size. The grinding wheel is attached to a mandrel and the mandrel is held in a grinding head. Two interchangeable grinding heads are generally supplied with each jig grinder. The function of the grinding head is to rotate the grinding wheel at a very high speed by means of an air-driven turbine. One grinding head operates at approximately 14,000 rpm and the other at 50,000 rpm.

In operation the grinding-wheel axis rotates around the axis of the main spindle while the grinding wheel is simultaneously rotating at a very high speed. In this movement it simulates a single-point boring tool. When grinding, it also traverses up and down inside the hole. The grinding

Courtesy of The Moore Special Tool Company

Fig. 15-32. Precision Jig Grinder. The table travel is 11 x 18 inches and the total accumulative accuracy (or error) is .000090 (90 millionths) inch.

wheel should never leave the hole entirely until it ceases to cut or is backed off from the side of the hole. Then it can be moved up, and out of the hole, in order to measure the hole or to move the table to another setting. The diameter of the hole is determined by how far the axis of the grinding wheel is offset from the axis of rotation of the main spindle. A coarse and a fine radial adjustment are available. The fine radial adjustment is made by a micrometer dial located above the spindle housing which is graduated to read in one-ten-thousandth (.0001) inch. The fine adjustment can be made while grinding. In addition to straight or cylindrical holes, it is possible to grind taper holes on this machine. This feature is invaluable when grinding draft angles on dies.

The operating procedures of a jig grinder are in some cases similar, and in other cases identical, to operating a jig borer. The principal difference is that grinding replaces boring in finishing the hole. The four basic steps in operating a jig grinder are: 1. align and clamp the workpiece on the jig-grinder table; 2. locate the reference axes of the workpiece with respect to the main spindle of the jig grinder; 3. locate the hole to be ground; and 4. grind the hole to size. In addition to grinding holes, the jig grinder is used to grind complicated contours on dies and on other tools.

Ch. 15 PRECISION HOLE LOCATION—THE JIG BORER

The jig borer and the jig grinder are able to machine holes to the most exacting standards to be encountered in industry. No two machines are more worthy of the attention of craftsmen. The precision work done on these machines can be personally satisfying and can engender a real pride in the craft of both the machinist and the toolmaker.

QUESTIONS

15-1. State the principle of the Transfer of Tolerances.

15-2. How can a hole be accurately located by boring?

15-3. What is the effect of boring through a hole that is not concentric with the axis of rotation of the boring tool? How can this be corrected?

15-4. The workpiece shown at A, Fig. 15-5, is to have several holes machined. These holes must be located very precisely in the part. What must be done to prepare the workpiece for this operation?

15-5. Describe how a precision layout is made using a vernier height-gage.

15-6. Describe how a precision layout is made using a universal precision or planer gage.

15-7. When and how are precision gage blocks used to make a precision layout?

15-8. What is a hand prick-punch? Why is it used? How is it used?

15-9. Why should a large drill not be used to start a hole from a prick-punch mark?

15-10. Why should a small drill be used to start a hole from a small prick-punch mark?

15-11. What is the sequence of operations for drilling a hole to precise locational tolerances from a precision layout?

15-12. What is a pump center and how is it used?

15-13. What is the sequence of operations for boring a hole to precise locational tolerances from a precision layout?

15-14. Describe how the trial-cut method is used to bore a hole in: a. a lathe; and b. a jig borer.

15-15. Describe three methods of attaching toolmakers buttons to a workpiece.

15-16. Describe two methods of locating toolmakers buttons on a workpiece.

15-17. How is a workpiece on which toolmakers buttons are attached located on the faceplate in the lathe?

15-18. How is the hole machined in the lathe after the toolmakers buttons have been removed?

15-19. Describe two methods of measuring the table movements used in the operation of jig borers.

15-20. What is the sequence of operations for machining accurate holes on a jig borer?

15-21. Why is it necessary to locate the workpiece accurately on the jig-borer table?

15-22. Describe three methods of picking up a reference edge of a workpiece on a jig borer.

15-23. What is the best sequence of metal-cutting operations for machining an accurate hole in a jig borer?

15-24. In a jig borer, why are holes sometimes finished by reaming?

15-25. When are holes finish machined by grinding?

15-26. Describe the motion of the grinding wheel, with respect to the main spindle and the work, in order to grind a true cylindrical hole.

15-27. The original dimensions and tolerances for conventional dimensions of the part illustrated are given in the accompanying table. The dimensions are to be converted to the coordinate system with sides N and P as the reference edges. All of the new tolerances must be equal-bilateral tolerances and they must be the maximum tolerances permissible.

Calculate the new dimensions and tolerances that replace those given in the chart in Columns I and II.

Basic Dimensions	Tolerances	
	I	II
A .5625	±.001	±.005
B .4375	±.005	+.000, − .004
C 1.0000	±.005	+.004, − .000
D 1.0000	±.005	±.004
E 1.8750	±.002	+.004, − .002
F 1.0000	±.004	+.004, − .002

15-28. The original dimensions and tolerances for the conventional dimensions of the part illustrated are given in the accompanying table. The dimensions are to be converted to the coordinate system with sides N and P as the reference edges. All of the new tolerances must be equal-bilateral tolerances and they must be the maximum permissible.

Ch. 15 PRECISION HOLE LOCATION—THE JIG BORER 363

Calculate the new dimensions and tolerances that replace those given in the chart under Columns I, II, and III.

Basic Dimensions	Tolerances		
	I	II	III
A 1.2500	±.002	±.005	±.004
B .6250	±.002	±.004	±.003
C 2.3750	±.005	±.010	+.000, − .005
D 2.5000	±.001	±.001	+.002, − .001
E 2.2500	±.005	±.001	+.003, − .002
F 1.2500	±.002	±.005	±.005
G 1.1250	±.002	±.005	±.003
H 1.7500	±.003	±.001	±.002
I 2.0000	±.004	±.001	±.002
J 1.0000	±.002	±.005	±.003
K 1.7500	±.002	±.003	±.004
L 1.8750	±.001	±.001	+.004, − .000
M 2.2500	±.005	±.003	±.004

CHAPTER **16**

Metal-Cutting Saws

Originally, metal-cutting saws were primarily used to cut off bar stock and structural steel, to length. This is still a very important area of application for metal-cutting saws; however, the range of application has been extended to include shaping of metal into finished and semifinished machine components, tools, dies, jigs, and fixtures. This work is called band machining. It has been made possible by the development of the metal-cutting band-saw blade along with the development of the vertical-contour band-sawing machine.

The Power Hacksaw

The power hacksaw is used to cut off bar stock, to length. The blade is held in a frame that is given a back and forth, or reciprocating, motion. The saw blade is forced into the work during the cutting stroke, while during the return stroke it is backed off from the work. This is done hydraulically on the saw shown in Fig. 16-1. The hydraulic pressure can be varied to suit the type of workpiece being cut. It has four cutting speeds (35, 70, 100, and 140 spm) which enable it to cut a variety of materials from aluminum to tough alloy steels. A coolant reservoir and a coolant pump located in the cabinet base provide an ample supply of cutting fluid at the saw blade. A fully automatic, power hacksaw is shown in Fig. 16-2. The operator merely loads the stock on the machine and sets it into operation by pushing the start button. After each cut has been completed, the traveling carriage, located on the stock rack, feeds the bar stock to the required position for the next cut. The hydraulically actuated vise then clamps the work and the saw begins to cut. An overhead clamp enables the work to be bundled, thereby making it possible to cut off two or more bars of stock during one cut. This machine can also be operated manually for short runs or single cuts.

Most power hacksaw blades are made from high-speed steel. Some blades have hardened high-speed steel teeth which are welded to a tough steel back. The size of the hacksaw blade teeth is dependent on the number of teeth per inch; fewer teeth per inch will result in a larger tooth. The number of teeth per inch to be used depends upon the hardness of the material being cut and upon the size and shape of the workpiece. The power hacksaw blade to be used in cutting mild materials in large sections should have 4 to 6 teeth per inch, while harder materials in large

Ch. 16 METAL-CUTTING SAWS 365

Courtesy of Racine Hydraulics & Machinery Inc.

Fig. 16-1. Power hacksawing machine.

sections should have 6 to 10 teeth per inch. When sawing pipes or tubing, the saw blade should have at least 14 teeth per inch. Structural steel shapes such as angles, T's, and channels require saw blades having 6 to

Courtesy of Racine Hydraulics & Machinery Inc.

Fig. 16-2. Automatic power hacksaw provided with overhead clamp to cut off bar stock in bundles.

14 teeth per inch depending on the thickness of the thinnest section and the way that the work is clamped in the hacksaw vise. Recommendations for cutting speed, teeth per inch, and feeding forces are given in Table 16-1.

Table 16-1. Recommended Saw-blade Pitch, Cutting Speed, and Feed for Power Hacksaw Operation

Material	Saw-blade Teeth, per inch	Cutting Speed, fpm	Feed Weight or Force, lbs.
Steel (125–180 Bhn)	4 to 6	120	150
Steel (181–250 Bhn)	6 to 10	100	125
Ferritic Stainless Steel	4 to 6	110	125
Austenitic Stainless Steel	4 to 6	80	125
Martensitic Stainless Steel	4 to 6	60	125
Carbon Tool Steel	6 to 10	75	125
High-Speed Steel	6 to 10	60	125
Structural Steel	6 to 10	120	125
Steel Tubing	14	120	60
Copper	6 to 10	130	125
Brass Tubing	14	130	60
Soft Brass	6 to 10	150	60
Hard Brass	6 to 10	110	60
Bronze	6 to 10	110	125
Nickel Silver	6 to 10	60	150
Manganese Bronze	6 to 10	80	60
Cast Iron	6 to 10	120	125
Iron Pipe	10 to 14	120	125
Aluminum Alloy	4 to 6	150	60
Aluminum–Heat Treated	4 to 10	120	60
Aluminum–Soft	4 to 6	180	60

Band Type Cutoff Saw

The band type cutoff saw, Fig. 16-3, utilizes a metal-cutting bandsaw blade to cut off bar stock and structural-steel sections, to length. The band-saw blades for these saws can be obtained already welded together into a loop of suitable size for each machine. The principle of operation of the band type cutoff saw is shown in Fig. 16-4. The saw blade revolves around the drive wheel and the idler wheel. The blade is twisted by the saw guides to bring the teeth into a vertical position in order to cut through the work. The guides also support the blade near the work, thereby reducing the tendency of the blade to bend away from the work. The blade can be made to cut at different speeds in order to cut different materials. One advantage of the band type cutoff saw is that the saw blade cuts continuously. Thus, the non-cutting time during the return stroke of the power hacksaw is saved. Furthermore, since it is not necessary to reverse the direction of the saw blade at the end of each stroke,

Ch. 16 METAL-CUTTING SAWS

Courtesy of the DoALL Company

Fig. 16-3. Band type cutoff sawing machine.

higher cutting speeds can be used on band type cutoff saws where the characteristics of the material being cut will allow their use. A continuously uniform feeding pressure is kept on the saw blade which is adjusted to suit the material being cut. Angular cuts can be taken on some band cutoff saws. Production band cutoff saws can be obtained that will handle bundles of stock and that will operate completely automatically.

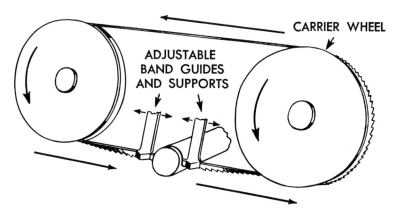

Courtesy of the DoALL Company

Fig. 16-4. Operating principle of the cutoff band saw.

Vertical-Contour Band-Sawing Machine

A vertical-contour band-sawing machine is illustrated in Fig. 16-5 with cabinet cover removed in one view, and all parts of the machine named. Although there are different makes and models of these machines their basic features are similar. The base supports the column and houses the hydraulic system for actuating the table feed, the electrical controls, the

Fig. 16-5. Vertical-contour band-sawing machine.

Ch. 16 METAL-CUTTING SAWS 369

electric-drive motor, the transmission, the drive wheel, and the air compressor. The column encloses the return route of the saw blade. It supports the head, and mounted on its face are the band-saw speed-tachometer, the blade shear, the table feed-rate and force control, the butt welder for welding blades, the grinder, and the on-off push button. The head of the machine houses the idler wheel and its adjusting mechanism, and it supports the saw guidepost.

Courtesy of the DoALL Company

(Left) Front view;(Right) Rear view.

370 METAL-CUTTING SAWS Ch. 16

The band-saw blade is driven by the lower drive wheel. The blade is supported by and runs the idler wheel located in the head. It must be held between these wheels with a predetermined amount of tension which is dependent upon the size of the blade and the material from which it is made. The tension is adjusted by turning the wheel-elevating screw. Many band saws have a tension indicator which is calibrated to read the recommended tension. The band must then be checked for tracking and alignment with the saw guides. The idler wheel in the head can be tilted to make this adjustment by releasing the lock knob and turning the adjusting knob (located in the center of this wheel), while turning the wheels slowly by hand. Some wheels are crowned, while others have a flange against which the blade runs. Crowned wheels are adjusted until the center of the blade runs on the crown of the wheel. On flanged wheels the blade is adjusted to ride against the flange. The blade must also be adjusted to the back-up thrust bearings which are located behind the saw guides, Fig. 16-6. The wheel-adjusting knob should be turned until the back edge of the saw blade just touches the back-up thrust bearing with no noticeable gap between them.

There are two sets of saw guides; one set is located below the table and the other set is attached to the guidepost above the table. In Fig. 16-6, the table has been removed to show both sets of saw guides. Saw guides are made in different widths to match the width of the saw blade. The guides should be adjusted until there is about .002-inch clearance between

Courtesy of the DoALL Company

Fig. 16-6. Precision, insert type saw guides and back-up thrust bearings. Table removed to show upper and lower guides.

the guide and the saw blade. The upper saw guide can be adjusted vertically. It should always be positioned as close to the work as possible in order to provide the maximum support to the saw blade. When sawing at high speeds (2,000 fpm, or more), roller type saw guides must be used.

Most vertical-contour band-sawing machines are equipped with an infinitely variable speed drive. The speeds are adjusted by simply turning a handwheel. The recommended cutting speeds for different materials are given in Table 16-2.

The saw table on most vertical band-sawing machines can be tilted 10- to 15-degrees to the left, and 45 degrees to the right. Some machines are equipped with a stationary table where the work must be fed into the saw blade by hand or by a weight-actuated feeding arrangement. The weight is attached to a cable system which pulls the work into the saw blade. Many contour band-sawing machines are equipped with a hydraulic table feed. The workpiece is clamped to the table and the table feed, actuated by a hydraulic cylinder, feeds the work into the saw blade. The hydraulic table feed has two characteristics, the *feed rate* and the *feeding pressure*, which are independent of each other and are adjusted separately. Their purpose should be understood.

The feed rate determines how fast the table will feed. It is controlled by a hydraulic flow-control valve which determines the rate of flow of the oil into the table feed-cylinder.

The feeding pressure determines the maximum force that can be exerted by the workpiece against the saw blade before the table will stall. Increasing the pressure setting will not increase the feed rate. The pressure setting does not determine the pressure that will be used to saw the workpiece. It will determine the pressure inside the table feed-cylinder at which the table will stall, as stated above.

The feed rate and the pressure that should be used depend upon many factors that, in combination, are unique for each job. Among the factors that must be considered in selecting the feed rate and pressure settings are: type of work material, thickness of the workpiece, type of saw blade, pitch of the saw blade, width and thickness of the saw blade, sharpness of the saw-blade teeth, surface finish required on the work, and the distance between the upper and lower saw guides.

Saw Blades

Most power hacksaw and band-saw blades are made from carbon alloy tool steel or from high-speed steel. Less commonly used are band-saw blades with small, cemented-carbide tips brazed on the end of the teeth. Band-saw blades are also made with industrial diamonds coating the edge of the blade or with either aluminum oxide or silicon-carbide abrasives coating the edges. The cemented-carbide tipped blades are used to saw difficult-to-machine materials such as: pure tungsten, beryllium, high-temperature alloys, foamed glass, etc. The abrasive-coated steel bands are used for deburring, corner breaking, and for rough and finish polishing of metals. They are also used to cut hardened steel, exotic

Table 16-2. Recommended Cutting Speeds for Band Sawing

Material	Carbon Tool-Steel, High-Speed Steel Work Thickness, inches						
	½	1	3	½	1	3	6+

Wait, let me redo this properly with correct columns.

Material	½	1	3	½	1	3	6+
	\multicolumn{7}{c}{Cutting Speed, fpm}						
Steel (AISI Numbers Given): 1108, 1132, 1212, 1213, 1137	200 to 225	175 to 200	140 to 175	275 to 350	225 to 275	200 to 225	150 to 200
Steel (AISI Numbers Given): 1151, 1008, 1013, 1020	150 to 175	125 to 150	100 to 125	250 to 300	200 to 250	150 to 200	125 to 150
Steel (AISI Numbers Given): 1035, 1320, 3115, 3130, 3135, 4017, 4130, 4317, 4608, 4621	125 to 150	100 to 125	75 to 100	200 to 250	175 to 200	125 to 175	90 to 125
Steel (AISI Numbers Given): 1040, 1060, 1345, 2330, 2345, 2512, 2517, 3150, 3310, 3315, 4068, 4042, 4047, 4140, 4150, 4340, 4350, 4640, 4812, 4820, 5045, 5120, 5135, 5140, 5168, 52100, 50100, 6117, 6120, 6145, 6152, 8615, 8715, 9255, 9260	100 to 125	80 to 100	50 to 80	175 to 225	140 to 175	115 to 140	75 to 115
Steel (AISI Numbers Given): 1090, 1095, 8645, 8715, 8647, 8660, 9310, 9317, 9347, 9445, 9447, 9763, 9840, 9850	80 to 100	50 to 80	50 to 70	200 to 225	150 to 200	125 to 150	90 to 125
Tool Steel: A-2, O-1, O-2, O-6	125 to 150	100 to 125	80 to 100	175 to 200	150 to 175	100 to 125	80 to 100
Tool Steel: M-1	100 to 125	75 to 100	50 to 75	130 to 160	100 to 130	75 to 100	50 to 75
Tool Steel: S-1, S-2, S-5, D-2, D-3, T-1, T-2, T-4, T-5, M-2, M-5	70 to 90	50 to 70	40 to 50	100 to 125	80 to 100	60 to 80	50 to 60
Tool Steel: H-12, H-13, H-21, H-22, H-24, H-25	60 to 75	50 to 60	40 to 50	160 to 200	125 to 150	100 to 120	80 to 90
Tool Steel: M-4, M-10, M-15	60 to 70	50 to 60	40 to 50	80 to 90	60 to 80	50 to 60	40 to 50

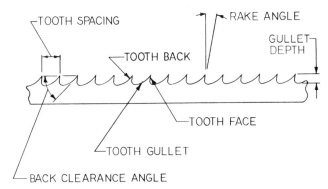

Fig. 16-7. Names of the parts of a saw blade.

alloys, and abrasive nonmetallic materials. Special steel blades are available for slicing such materials as cork and foam rubber.

The majority of the work on band saws, however, is done with carbon-alloy steel or high-speed steel saw blades. The recommended cutting speeds given in Table 16-2 are for sawing different grades of steel with carbon-tool steel or high-speed steel saw blades. These recommendations should only be used as a starting point. Conditions at the job can make deviations from these recommendations necessary.

The names of the parts of a saw blade are given in Fig. 16-7. There are four, basic tooth forms which are shown in Fig. 16-8. The precision, or regular-tooth, form is the most common. It has a zero-degree rake angle and a 30-degree back-clearance angle. The gullet is deep and has a

Courtesy of the DoALL Company

Fig. 16-8. The basic band-saw blade-tooth forms. The names of the tooth forms from top to bottom are: buttress, claw tooth, precision, and tungsten carbide.

smooth radius at the bottom providing an ample chip-capacity for most sawing operations. It cuts accurately and can produce a good surface finish. The precision-tooth form is the most versatile and is recommended for contour work and for straight cuts. The buttress, or skip tooth, tooth form is the same as the precision-tooth form given except that the gullet has been lengthened to handle larger chip loads. It is primarily recommended for taking deep cuts in nonferrous metals, wood, and plastics. The claw-tooth form has a positive rake angle and a lengthened gullet similar to the buttress-tooth form. It is recommended for cutting all metals, wood, and plastics and is particularly recommended for thick work-sections. The positive rake angle requires a minimum feed pressure to penetrate into the work, thereby extending the tool life of the blade. The carbide blade has small carbide tips brazed on the end which are used for cutting difficult-to-cut materials.

There are five characteristics that must be specified in order to select a saw blade. These characteristics are: the tooth form, pitch, width, gage, and set. The pitch is the number of teeth per inch. The pitch is selected primarily on the basis of the thickness of the workpiece to be cut. At least two teeth should always be in contact with the work; however, it is preferable to have even more teeth in contact. Other factors, such as work material and the finish desired must also be considered in making the final selection of the pitch. For average work conditions the following pitches are recommended: ¼- to ½-inch thick, 8-10 pitch; ½- to 1-inch thick, 8 pitch; 1- to 3-inches thick, 6 pitch; 3- to 6-inches thick, 4 pitch; 6 inches and over, 3 pitch claw-tooth form. The other pitches recommended should have the precision-tooth form. The width of the blade is the distance from the tip of the tooth to the back of the blade. The width of the blade is selected on the basis of the minimum radius that is to be sawed. A radius chart for selecting the blade width is shown in Fig. 16-9. A wide blade is most accurate for cutting a straight line. The gage of the blade is the thickness of the band expressed in inches. A thicker blade has

Courtesy of the DoALL Company

Fig. 16-9. Radius chart for selecting band-saw width.

Ch. 16 METAL-CUTTING SAWS 375

Courtesy of the DoALL Company

Fig. 16-10. Basic band-saw tooth patterns. (Left) Wavy set. (Center) Straight set. (Right) Raker set.

a greater beam strength, or resistance to bending. When the width of the blade is limited by the machine capacity or by the radius to be cut, the beam strength can be increased by using the same width blade but in a heavier gage. The heavier gages also have more resistance to side displacement and are recommended for sawing large workpieces where extreme accuracy is required. Set is the total amount that the teeth are offset (Fig. 16-10), to provide side clearance for the back of the band. The set of the saw blade enables it to cut contours. The width of the kerf, or slot, cut by

Courtesy of the DoALL Company

Fig. 16-11. Grinding end of band-saw blade in preparation for welding. Note saw teeth on opposite sides.

Fig. 16-12. Preparation of teeth prior to welding, to insure correct tooth-spacing on welded blade.

the blade is greater than the gage which allows the back of the saw blade to turn. There are three types of set patterns; raker, straight, and wave. The raker set is the most common. It has one tooth offset to the right, one to the left, and the next one remains straight. It is recommended for work of constant cross section, from thin to extra heavy, such as bar stock, forgings, and die blocks. The wave set has a group of teeth offset to the left and another group offset to the right to form a wave pattern. It is recommended for work having a varying cross section, such as pipe, tubing, angles, and structural sections. The straight set has every other tooth alternating from left to right. It is used for free-cutting materials such as aluminum or wood.

The saw blade used on vertical-contour band-sawing machines is obtained in coils and must be welded together to form a loop. The band

may be cut to length on the blade shear (which is attached to the machine) or with tin snips. When snips are used, the ends of the blade must be ground to make certain that they will fit together perfectly. The ends are brought together with the teeth on opposite sides, as shown in Fig. 16-11, and the ends are ground. It is not necessary to grind the ends perfectly square. When the blade is ground with the ends held as in Fig. 16-11 they will match perfectly when brought together to form the loop regardless of the angle to which they are ground. Some of the teeth must be ground off from all blades having a pitch from 4 to 10 so that the tooth spacing will be uniform after the weld has been made. About $\frac{1}{4}$ inch of metal will be consumed during the welding process and this loss of material must be taken into account. Instructions for grinding are given in Fig. 16-12. This grinding operation must be done with great care in order to grind the right number of teeth, or portion of a tooth, and not to grind below the gullet which would weaken the blade.

After the blade has been properly prepared for welding it is inserted in the butt welder, Fig. 16-13, with the teeth facing out and with the back of the blade firmly seated against the back alignment surface of the welder. One jaw of the welder is adjusted to bring a pressure to bear on the two ends of the blade that are in contact. The amount of this pressure and the voltage setting of the welder are obtained on charts that are furnished with each machine. If the saw blade is to be used to make an inside cut, as in Fig. 16-13, make certain that the blade passes through the hole in the workpiece before the weld is made. The weld is made auto-

Courtesy of the DoALL Company

Fig. 16-13. Welding a band-saw blade together in preparation for making an inside cut on the part.

Courtesy of the DoALL Company

Fig. 16-14. Example of a perfect band-saw-blade weld.

matically when the welding switch is pressed. The weld is made by passing a low-voltage high-amperage current through the welder jaws and into the blade. The current flows toward the ends of the blade where the resistance to the current flow causes the metal in this area to heat up. When this happens, the metal in the joint becomes soft and flows together. The movable jaw of the welder causes the metal in the weld to compress together and form a flash. When the weld is completed it will be very hard and very brittle. Before it can be used it must be annealed and the flash must be removed. The welding flash can be removed on the grinder. Care should be exercised to avoid grinding the blade too thin and thereby reducing its strength. The weld is annealed by clamping the blade in the jaws of the butt welder. The welder is adjusted to anneal and heat is applied by pushing in or jogging the annealing switch until the weld reaches a dull cherry-red color. Avoid overheating the weld as this could well re-harden it. Before the blade is placed in the machine it should be inspected visually. One example of a perfect weld is shown in Fig. 16-14.

Band Machining

The basic approach to band machining is similar to working on other machine tools. The following steps should be taken: 1. analyze the job; 2. make a layout; 3. set up the machine; 4. cut the work to size on the machine; and 5. inspect the work. The job is begun by analyzing how it can be done most effectively. The next step is to make a layout. In the case of band machining this must be done carefully and accurately because the final results can be no better than the layout. The machine setup involves installing the correct blade, and setting the machine to cut at the correct cutting speed. If a power feed is to be used the workpiece will have to be held in place or clamped to the table, and the correct feed rate and pressure settings made. The part is cut by feeding the workpiece past the blade and keeping as close as possible to the layout line.

Finally, the work is inspected in order to make certain that the results are as specified by the part drawing.

The workpiece is frequently fed into the cut by hand. It can be held directly in the hand as shown in Fig. 16-15. This illustration shows a blanking punch and die, being made in one operation. The job is started by first laying out the die. A hole is drilled inside the layout of the die at an angle (Fig. 16-16), so that it intersects the location of the layout line at the center of the two-inch thickness and comes through the die block on the other side, at a location that is outside the layout line of the punch. A narrow saw blade is passed through the hole and welded together. The saw table is set at the required angle for the saw cut (7 degrees, in this case) and the punch is cut out by sawing just inside the die layout-line and just outside the punch layout-line (Fig. 16-16). Thus, there is enough material on the punch and the die so that they can be filed to fit together with the required amount of clearance. Several punches and dies finished by this method are shown in Fig. 16-17. The filing of the punch and the die can also be done on the vertical-contour band saw, using a file band. This is a quick and economical method of making a short-run trimming, blanking or shaving die. It saves material because the slug that is removed from the center is made into the punch. If this method of making a punch and a die from one part is to be used, the part cannot have any sharp corners or radii. In other words, the saw must be able to cut continuously around the die without interruption.

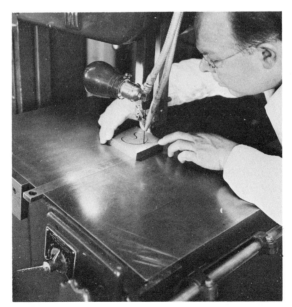

Courtesy of the DoALL Company

Fig. 16-15. Sawing a blanking punch and die in one operation, with the work table tilted.

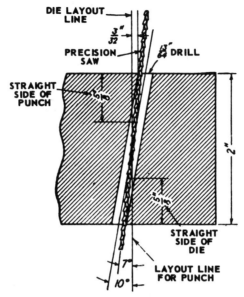

Courtesy of the DoALL Company

Fig. 16-16. Diagram showing the angles of the starting hole and the saw band in 2-inch material when making a blanking punch and die in one operation.

Courtesy of the DoALL Company

Fig. 16-17. Several finished punches and dies made on the vertical-contour band saw. Note remainder of starting hole on punch in upper-right view.

The dimensions necessary to design and make a die by this method are given in Table 16-3.

Table 16-3. Basic Dimensions for Making Punch and Die, by Method Shown in Fig. 16-16

Die Thickness, inches	Angle of Saw Starting Hole, degrees	Angle for Saw Cut, degrees	Distance from Die Layout Line to Center of Saw Kerf, inches	Distance from Die Layout Line to Center of Starting Hole, inches	Diameter of Drill, inches	Width of Starting Saw, inches	Amount of Straight Sides on Punch and Die
$1/2$	21	18	$5/64$	$3/32$	$1/8$	$3/32$	$3/16$
$3/4$	18	15	$3/32$	$1/8$	$1/8$	$3/32$	$9/32$
1	14	11	$3/32$	$1/8$	$9/64$	$1/8$	$3/8$
$1\ 1/4$	12	9	$3/32$	$1/8$	$9/64$	$1/8$	$15/32$
$1\ 1/2$	11	8	$7/64$	$1/8$	$9/64$	$1/8$	$9/16$
2	10	7	$1/8$	$3/16$	$13/64$	$3/16$	$13/16$
3	9	6	$5/32$	$1/4$	$17/64$	$1/4$	$1\ 1/8$
4	8	6	$7/32$	$9/32$	$17/64$	$1/4$	$1\ 5/8$
5	7	6	$1/4$	$5/16$	$17/64$	$1/4$	$2\ 1/4$
6	6	5	$1/4$	$5/16$	$17/64$	$1/4$	$2\ 1/2$

Die Thickness, inches	Minimum Outside Radius, inches	Minimum Inside Radius, inches	Saw Width, inches
$1/2$	$9/32$	$1/8$	$3/32$
$3/4$	$5/16$	$1/8$	$3/32$
1	$7/16$	$1/4$	$1/8$
$1\ 1/4$	$7/16$	$1/4$	$1/8$
$1\ 1/2$	$15/32$	$1/4$	$1/8$
2	$11/16$	$7/16$	$3/16$
3	$7/8$	$9/16$	$1/4$
4	$1\ 1/32$	$9/16$	$1/4$
5	$1\ 1/16$	$9/16$	$1/4$
6	$1\ 1/8$	$9/16$	$1/4$

Courtesy of the DoAll Company

Another fast method of making low-cost dies is called the Continental Process. This process is illustrated in Figs. 16-18 through 16-21. The die, punch, and a guide plate are made from thin sections of steel, on a vertical-contour band saw. To make the stamping, Fig. 16-18, the work material, which is a piece of sheet metal, is placed between the die and the guide plate. The punch is then placed in the guide plate, over the work. The ram of a punch press (or arbor press) forces the punch into the die, thereby cutting out the blank. Dies for drawing, bending, and a variety of other stamping operations can be made by this process.

In this case, the punch and the die are made separately. The first step is to carefully lay out the die. Four holes are drilled in the corners of the die block to receive dowel pins which are used to hold the guide plate. Corresponding holes are drilled in the guide plate using the holes in the die block to guide the drill. After the punch plate is assembled tightly against the die block, a hole is drilled in this assembly, inside of the lay-out lines. The band-saw blade is inserted through this hole (see Fig.

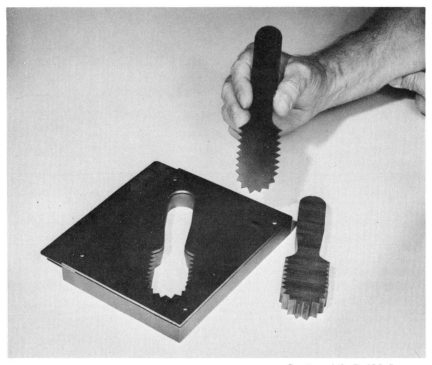

Courtesy of the DoALL Company

Fig. 16-18. Blanking punch and die made by the Continental Process on a vertical-contour band-sawing machine.

16-13) and welded together. The opening is then sawed out by manually guiding the work so that the saw cuts just inside the layout line allowing a small amount of metal to remain for filing, Fig. 16-19. The workpiece is held in a workjaw which makes it easier to guide the workpiece and follow the layout. After cutting, a band file (Fig. 16-20) is placed in the saw and the die block and guide plate are filed to size. The punch plate is disassembled from the die block and is used to lay out the punch. It is then reassembled to the die block with suitable shims between them. The shims should be just large enough to allow the work material to be inserted. The punch is then cut and filed to size on the band saw (see Fig. 16-21).

The cutting of contours can be facilitated by the attachment shown in Fig. 16-22. This attachment is called a servo-contour directional feed control. The operator follows the contour on the part by regulating the hydraulic table-feed rate during the cut and by turning the handwheel which rotates the parts through a movement of the roller chain. The roller chain can be placed directly around the work, as shown, or the work can be placed in a workjaw (Fig. 16-19) which would, in this case,

have the roller chain placed around it. Three-dimensional parts such as those shown in Fig. 16-23 can be made on the vertical-contour band-sawing machine. This machine can be provided with fixtures enabling it to do production work. An example is seen in Fig. 16-24 where the cap of a large Diesel engine connecting-rod forging is being separated. The fixture firmly holds the forging and accurately locates it with respect to the saw blade. The fixture is clamped to the table so that the cut is made using the power table-feed.

A great variety of additional work can be done on the vertical-contour band saw. Friction sawing, which utilizes the friction between the saw blade and the work to generate high temperatures a short distance ahead of the blade in the workpiece, is capable of cutting at extremely high feed rates. Metals that are impossible to cut in any other way can be cut by this method. Also, thin aircraft materials, distorted when sawed by conventional methods, can be sawed safely by friction sawing. The cutting speed of the saw blade for friction sawing is between 6,000 and 15,000 fpm. Friction sawing is possible because each saw tooth is in contact with the work for only a fraction of a second and there is too little time

Courtesy of the DoALL Company

Fig. 16-19. Sawing contour of die to layout line using work jaw to feed the workpiece.

384 METAL-CUTTING SAWS Ch. 16

Courtesy of the DoALL Company

Fig. 16-20. Filing the block of the die using a file band.

Courtesy of the DoALL Company

Fig. 16-21. Sawing the contour of a punch using a flat vise to hold and feed the workpiece.

Ch. 16 METAL-CUTTING SAWS 385

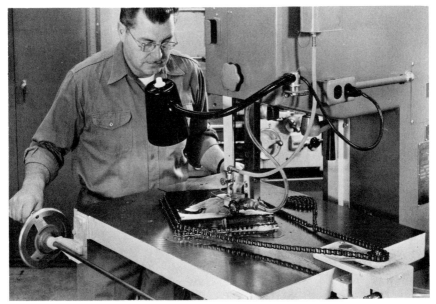

Courtesy of the DoALL Company

Fig. 16-22. Combining the hydraulic table-feed and servo-contour directional feed-control for cutting straight lines and contours.

Courtesy of the DoALL Company

Fig. 16-23. Combining various shaping cuts in several planes enables the vertical-contour band saw to produce very complicated parts.

Courtesy of the DoALL Company

Fig. 16-24. Separating the cap of a large, connecting-rod forging which is held in a fixture.

Courtesy of The L. S. Starrett Company

Fig. 16-25. Hole saw.

for heat to transfer into the tooth. Certain materials such as sponge rubber, cloth, cork, leather, synthetic sponges, and corrugated aluminum are cut by slicing with a continuous knife blade that is revolved on the wheels of the saw. Intricate scroll work can be done with spiral saw bands. Grinding and polishing blades are used to cut certain materials, to deburr, and to break sharp corners. Electro-band machining is done by feeding a low-voltage high-amperage current into the knife band at a rate that causes it to discharge a sustained arc from the knife edge onto the material being cut. A flood of coolant quenches the arc to prevent it from burning the material. The material is disintegrated by the arc without any actual contact between the work and the band. This band-machining process is used to cut heat-exchanger cores, honeycomb structures, flexible tubing, etc. The method requires a special machine and cannot be adapted to the ordinary vertical-contour band saw.

Hole Saws

Hole saws are used to cut holes in sheet metal, tubing, and in other thin metal parts. The thrust force of the larger twist drills would deflect and bend, rather than penetrate, these parts. The hole saw, Fig. 16-25, has a small twist drill in the center which drills a small hole before the saw comes in contact with the work. The small twist drill then acts as a

Courtesy of The L. S. Starrett Company

Fig. 16-26. Drilling a hole in tubing with a hole saw.

pilot to locate and to steady the saw while it cuts through the metal to produce a larger hole. Figure 16-26 illustrates a hole being drilled in tubing with a hole saw.

QUESTIONS

16-1. What type of holding device is required to cut off a bundle of bar stock simultaneously?

16.2 What method is used to orient the blade of a band type cutoff saw toward the workpiece?

16-3. Describe the procedure for making a band-saw blade track on the carrier wheels of the saw.

16-4. What is the function of the feed rate adjustment and the pressure adjustment for the hydraulic-power table feed?

16-5. From what materials are band-saw blades made?

16-6. What are the following parts of saw teeth?
 a. rake angle d. gullet
 b. tooth back e. tooth facing
 c. tooth face f. pitch

16-7. Describe each of the four, basic band-saw tooth forms.

16-8. What five characteristics must be described in order to select a band-saw blade?

16-9. Describe the three types of set patterns that are used on band-saw blades.

16-10. For what application should the wave set pattern be used?

16-11. Describe the procedure required to prepare a band-saw blade for welding.

16-12. What basic steps should be taken to do band machining?

16-13. Describe how a punch and die can be made from a single piece of metal on a contour band saw.

16-14. Describe how the punch, the die, and the guide plate are made for a Continental Process die.

16-15. What is friction sawing?

Appendix

Table A-1. Recommended Cutting Speeds for Drilling with High-Speed Steel Drills. For Reamers Use ½ to ⅔ Speed Given in this Table

Material	Hardness, Bhn	Cutting Speed, fpm	Material	Hardness, Bhn	Cutting Speed, fpm
Plain Carbon Steels			*Stainless Steels (Cont.)*		
AISI-1019, 1020,	120–150	80–120	Cold-Drawn	225–275	60–90
1030, 1040, 1050,	150–170	70–90	Ferritic	135–185	100–120
1060, 1070, 1080, 1090	170–190	60–80	Martensitic		
	190–220	50–70	Annealed	135–185	100–130
	220–280	40–50	Cold-Drawn	185–240	90–120
	280–350	30–40	Quenched & Tempered	275–325	50–60
	350–425	15–30		375–425	30–40
Alloy Steels			*Tool Steels*		
AISI-1320, 2317, 2515,	125–175	60–80	Water Hardening	150–250	70–80
3120, 3316, 4012, 4020,	175–225	50–70	Cold Work	200–250	20–40
4120, 4128, 4320, 4620,	225–275	45–60	Shock Resisting	175–225	40–50
4720, 4820, 5020, 5120,	275–325	35–55	Mold	100–150	60–70
6120, 6325, 6415, 8620,	325–375	30–40		150–200	50–60
8720, 9315	375–425	15–30	High-Speed Steel	200–250	30–40
				250–275	15–30
Alloy Steels					
AISI-1330, 1340, 2330,	175–225	50–70	*Gray Cast-Iron*		
2340, 3130, 3140, 3150,	225–275	40–60		110–140	90–140
4030, 4063, 4130, 4140,	275–325	30–50		150–190	80–100
4150, 4340, 4640, 5130,	325–375	25–40		190–220	60–80
5140, 5160, 52100, 6150,	375–425	15–30		220–260	50–70
6180, 6240, 6290, 6340,				260–320	30–40
6380, 8640, 8660, 8740,					
9260, 9445, 9840, 9850			*Malleable Iron*		
			Ferritic	110–160	120–140
Stainless Steels			Pearlitic	160–200	90–110
Standard Grades				200–240	60–90
Austenitic				240–280	50–60
Annealed	135–185	40–50			
Cold-Drawn	225–275	30–40	*Aluminum Alloys*		
Ferritic	135–185	50–60	Cast–Non Heat Treated		200–300
Martensitic			Cast–Heat Treated		150–250
Annealed	135–175	55–70	Wrought–Cold Drawn		150–300
	175–225	50–60	Wrought–Heat Treated		140–300
Quenched & Tempered	275–325	30–40			
	375–425	15–30	*Brass & Bronze (Ordinary)*		150–300
Free Machining Grades			*Bronze (High Strength)*		
Austenitic					
Annealed	135–185	80–100			30–100

Table A-2. Recommended Feed Rate for High-Speed-Steel Twist Drills

Drill Diameter, Inches	Feed, Inches per Revolution	Drill Diameter, Inches	Feed, Inches per Revolution	Drill Diameter, Inches	Feed, Inches per Revolution
1/16 to 1/8	.001 to .003	1/4 to 1/2	.004 to .010	over 1	.015 to .025
1/8 to 1/4	.002 to .006	1/2 to 1	.007 to .015		

Table A-3. Cutting Speeds for Machine Tapping

Material	Cutting Speed, fpm	Material	Cutting Speed, fpm
Low Carbon Steels Up to .25% C	40 to 80	*Aluminum*	50 to 200
		Brass	50 to 200
Medium Carbon Steels .30 to .60% C		*Manganese Bronze*	30 to 60
Annealed	30 to 60	*Phosphor Bronze*	30 to 60
Heat Treated (220 to 280 Bhn)	20 to 50	*Naval Brass*	80 to 100
Tool Steels, High Carbon & High-Speed Steel	20 to 40	*Monel Metal*	20 to 40
		Tobin Bronze	80 to 100
Stainless Steels	5 to 35	*Plastics*	
Gray Cast-Iron	40 to 100	Thermoplastics	50 to 100
		Thermosetting	50 to 100
Malleable Iron		*Hard Rubber*	50 to 100
Ferritic	80 to 120		
Pearlitic	40 to 80	*Bakelite*	50 to 100
Zinc Die Castings	60 to 150		

Table A-4. Decimal Equivalents

		1/64	.015625			33/64	.515625
	1/32		.03125		17/32		.53125
		3/64	.046875			35/64	.546875
1/16			.0625	9/16			.5625
		5/64	.078125			37/64	.578125
	3/32		.09375		19/32		.59375
		7/64	.109375			39/64	.609375
1/8			.125	5/8			.625
		9/64	.140625			41/64	.640625
	5/32		.15625		21/32		.65625
		11/64	.171875			43/64	.671875
3/16			.1875	11/16			.6875
		13/64	.203125			45/64	.703125
	7/32		.21875		23/32		.71875
		15/64	.234375			47/64	.734375
1/4			.250	3/4			.750
		17/64	.265625			49/64	.765625
	9/32		.28125		25/32		.78125
		19/64	.296875			51/64	.796875
5/16			.3125	13/16			.8125
		21/64	.328125			53/64	.828125
	11/32		.34375		27/32		.84375
		23/64	.359375			55/64	.859375
3/8			.375	7/8			.875
		25/64	.390625			57/64	.890625
	13/32		.40625		29/32		.90625
		27/64	.421875			59/64	.921875
7/16			.4375	15/16			.9375
		29/64	.453125			61/64	.953125
	15/32		.46875		31/32		.96875
		31/64	.484375			63/64	.984375
1/2			.500	1			1.000

Index

Accuracy, 41
Acme thread cutting on an engine lathe, 297
Allowance, 40
American National Thread System, 255
American Standard Acme Threads, 256–258
American Standard Buttress Screw Thread, 259
American Standard Machine Tapers, 221
American Standard Microscope Objective Threads, 256
American Standard Taper Pipe Threads, 258, 259
American Standard thread cutting on an engine lathe, 289–293
American Standard Unified Miniature Screw Threads, 256
American Standard Unified Threads, 250–255
Angle plate, 45, 96, 236, 240–246
Angle turning, 230–232

Band machining, 378–387
Band sawing, servo-directional feed control for contour, 382, 383
Band saw, vertical contour, 368–371
 feed rate for, 372, 373
 saw blades for, 371–378
 welding blades for, 376–378
Band type cutoff saw, 366, 367
Basic size, 41
Bevel protractor, vernier, 25–27
Bilateral tolerance, 41
Borer, jig, 341–344
Boring head, offset, 329, 353, 354
Boring, internal tapers and angles, 232–234
 internal threads, 294–297
 objectives of, 215
 on engine lathes, 215–217, 336–340
 on jig borers, 351–355
 on turret lathes, 311

Boring, principles of, 7, 215, 328–330
 taper, 232–234
Boring mill, vertical, 304, 318–321
Brinell hardness, 127
British Association Standard (BA) threads, 260
British Standard Fine (BSF) threads, 259, 260
British Standard I.S.O. Metric threads, 260, 261
British Standard Whitworth threads, 259, 260
Broaching, principles of, 8, 9
Brown & Sharpe Taper, 220
Built-up edge, 160
Buttoning as a hole location method, 337–341, 356–359

Calipers, 16–19
 gear tooth (vernier), 279
 hermaphrodite, 16, 17
 inside, 16–19
 micrometer, 27–32, 272, 273
 outside, 16, 17
 vernier, 25
Carbon tool steel as a cutting tool material, 137
Cast nonferrous alloys as cutting tool materials, 139
Cemented carbides as cutting tool materials, 138
Center drill, 169
Center drilling for starting a drilled hole, 100, 101, 211, 311, 312, 336, 340, 352, 353
Center drilling in preparing work for cylindrical turning, 169–173
Center gage, 175, 278
Center head, 19, 20
Centering microscope, 349
Center, lathe, 174–179
 aligning, 176–179
Center turning, 180–186

394 INDEX

Ceramic cutting tools, 138, 139
Change gear ratio formula for thread cutting, 281
Chatter, 79, 144, 145
Chip breakers, 161
Chip formation, 157–160
 in drilling, 75, 77
Chipping, cutting edge, 164, 165
Chucking part, in collet chuck, 204–206
 in four-jaw chuck, 201, 204
 in three-jaw chuck, 200, 201
Chucks, collet, 121, 204–209
 drill, 122
 four-jaw independent, 122, 201–204
 three-jaw universal, 121, 122, 200, 201
Collet chuck, 121, 204–209
Combination drill and countersink, 169
Combination square, 19
Combination stock stop and start drill, 311, 312
Comparators, optical, 266–268
Compound rest method of taper turning, 230–232
Compound rest setting for engine lathe thread cutting, 275–301
Continental Process for making dies, 381, 382
Contour band sawing, servo-directional feed control for, 382, 383
Coordinate dimensions, 323–328
Coordinate measuring system, 341–344
Core drill, 311
Counterbores, 80
Counterboring, principles of, 8
Countersink, 80
 combination drill and, 169
Crater wear, 163
Cutter holders, turret lathe, 310
Cutting edge, chipping, 164, 165
 honing, 165
Cutting fluids, for preventing formation of built-up edge, 160
 for reaming, 102
 for tapping, 104
 for turning, 132, 188
Cutting force, 141–144
Cutting-off, 217–219, 315
Cutting speeds, for drilling, 90–92
 for power hacksawing, 366
 for reaming, 79
 for tapping, 102–104
 for thread cutting, 285, 286
 for turning, 128–136
 for vertical contour bandsawing, 372, 373
Cutting tool materials, 136–139
 carbon steel, 137

Cutting tool materials, cast nonferrous alloys, 139
 cemented carbides, 138
 ceramic, 138, 139
 high-speed steel, 137
 industrial diamonds, 139
Cutting tools, general characteristics of, 3, 4
 grinding for single-point, 153–157
 materials for, 136–139
 nomenclature of single-point, 139–146
 shapes of single-point, 148–153
 thread, 275–280
 wear of, 161–165
Cylindrical grinding, 11
Cylindrical square, 21

Decimal-inch system, 14–16
Depth gage, vernier, 25
Dial test indicators, 33, 34
Diamond cutting tools, 139
Die head, self-opening, 313
Digital readout system, 186
Dividers, 46, 50
Drill, center, 169
 chuck, 122
 combination drill and countersink, 169
 core, 311
 cutting action, 70–73
 elements of twist, 67–70
 gun, 84, 85
 horsepower, 85–87
 point angles, 70
 presses (see *Drilling machines*)
 relief angles, 69, 70
 sizes, 77
 spade, 83, 84, 312
 spot, 352
 start, 311
 thrust force, 70, 76, 85, 86
 twist, 67–70
 web thinning, 70–73
Drilling, basic principles of, 7, 8
 chip formation in, 75, 77
 feed rate in, 92
 on a drill press, 99–101
 on a lathe, 211–214
 on a turret lathe, 310–312
 oversize amount normally cut, 177
 setups for, 92–99
 speed, 90–92
Drilling machines, 61–67
 gang, 66, 67
 multiple spindle, 66, 67
 radial, 64, 65
 sensitive, 61–64

INDEX 395

Drilling machines, turret, 65, 66
 upright, 61
Drill presses (see *Drilling machines*)
Driver plate, 180

Edge build-up, 160
Edge finder, 349–351
Electro-band machining, 387
End-cutting-edge angle, 143
Engine lathe, 108–121
 apron, 119, 120
 back gears, 113
 bed, 109–111
 boring on, 215–217, 336–340
 carriage, 116
 compound rest, 119
 cross slide, 116–118
 feed-thread-gear box, 116
 headstock, 111–113
 headstock spindle, 114–116
 headstock spindle bearings, 114
 micrometer dial, 116–119
 spindle nose, 114–116
 tailstock, 120, 121
 thread cutting on an, 275–301 (see *Thread cutting on an engine lathe*)
 ways, 111

Faceplate, 122, 236, 237–247
 chuck, 247
 jaws, 237, 238
 setups, 238–247
 work-precision hole location, 336–340, 356–359
Facing, end of work held between centers, 182, 183
 work held in chuck, 210
Feed rate, drilling, 92
 power hacksawing, 366
 reaming, 79
 turning, 132–135, 182, 197, 198
 vertical contour bandsawing, 372, 373
Filing on a lathe, 186
Finish turning, 168
Fixtures, 105, 106
Flank wear, 163
Formulas, change gear ratio in thread cutting, 281
 chord of holes equally spaced on a circle, 51
 cutting speed in drilling, 91
 cutting speed in turning, 135
 drilling thrust, 85
 horsepower for drilling, 87
 horsepower for turning, 195, 196
 sine bar, 224
 taper calculations, 221, 222, 225

Formulas, taper turning (offset center method), 230
 thread cutting of metric threads, 283
 thread cutting of multiple threads, 299, 300
 thread cutting of worm threads, 284
 thread pitch, 248
 three wire thread measurement, 269, 270
Fractional-inch system, 14–16
Friction sawing, 383–387

Gage blocks, precision, 34–38
Gages, center, 175, 278
 height (vernier), 25
 planer, 39, 40
 small hole, 19
 surface, 38, 39
 taper plug and ring, 222, 223
 telescoping, 19
 thread plug and ring, 262–264
 thread snap, 265, 266
 universal precision, 39, 40
Gaging and measuring screw threads, 262–273
Gang drilling machine, 66, 67
Gear tooth caliper, vernier, 279
Grinder, jig, 359, 360
Grinding attachment, tool post, 175
Grinding, cylindrical, 11
 jig, 356–360
 precision, 10–12
 single-point tool, 153–157
 surface, 12
Gun drill, 84, 85

Hacksaw, power operated, 364
 feed rate for, 366
 saw blades for, 364–366
Hand reamers, 79
Hardness, Brinell, 127
 Rockwell, 127
Hardness testing, 127, 128
Height gage, vernier, 25
Hermaphrodite calipers, 16, 17
High-speed steel as a cutting tool material, 137
Hole gages, small, 19
Hole location methods, buttoning, 337–341, 356–359
 jig boring, 344–355
 jig grinding, 356–360
 layout and drilling, 330–335
 layout, drilling and boring, 336, 337
Hole saws, 387, 388
Holes equally spaced on a circle, formula for chords for, 51

396 INDEX

Honing, cutting edge, 165
Horsepower, at the cutting tool, 195
 at the motor, 195, 196
 drilling, 85–87
 turning, 194–198
 unit, 194, 195

Indicators, dial test, 33, 34
Inserts, single-point, disposable or throwaway, 147, 148, 151, 152
Interference fit screw threads, 256
Internal threads, boring, 294–297
International Metric Thread System, 261, 262

Jarno Taper, 220
Jig borer, 341–344
Jig boring practice, 344–355
 aligning and clamping the workpiece, 344–347
 as a hole location method, 344–355
 drilling and boring hole, 351–355
 locating the hole, 351
 locating workpiece and spindle, 347–351
Jig grinder, 359, 360
Jig grinding, 356–360
 as a hole location method, 356–360
Jigs, 105, 106

Knurling, 193, 194

Lathe attachment, collet chuck, 121, 204–209
 drill chuck, 122
 faceplate, 122, 236, 237
 follower rest, 192, 193
 four-jaw independent chuck, 122, 201–204
 micrometer carriage stop, 122
 steady rest, 122, 123, 172, 191, 192
 taper attachment, 226
 three-jaw universal chuck, 121, 122, 200, 201
 tool post grinding attachment, 175
 tracer attachment, 123–125
Lathe centers, 174–179
 aligning, 176–179
Lathes, boring on, 215–217, 336–340
 components for engine, 109–121 (see *Engine lathe*)
 engine, 108–121
 filing on, 186
 method of leveling, 173, 174
 production, 317, 318
 turret, 304–309
 vertical, 304, 318–321

Layout and drilling as a hole location method, 330–335
Layout, drilling, and boring as a hole location method, 336, 337
Layout machines, 59
Layout procedures, 44–58
 for cylindrical part, 47–49
 for equally spaced holes on a circle, 49–51
 for hole location, 45–47
 for large castings, 51–58
 painting, 44, 45
Layout scale, 54, 55
Lead angle, 143–145, 165
Left-hand thread cutting on an engine lathe, 293, 295
Leveling lathes, method of, 173, 174
Limits, 40
 lower limit, 40
 upper limit, 40

Machine reamers, 77–79
Machine shop, the work of the, 1
Machine tools, general description, 2, 3
Machining, band, 378–387
 electro-band, 387
 turret lathe, 309–317
Mandrels, 189–191
Master precision square, 20
Measuring and gaging tapers, 222–226
Measuring system, coordinate, 341–344
Mechanical efficiency factor, 196
Metric thread cutting on an engine lathe, 283, 284
Micrometer measuring instruments, 27–32, 272, 273
 depth micrometer, 32
 inside micrometer, 32
 outside micrometer caliper, 27–32
 screw thread micrometer, 272, 273
Micrometer scale, reading, 29–32
Microscope, centering, 349
Milling, 5–7
Milling Machine Taper, 221
Morse Taper, 220
Multiple spindle drilling machine, 66, 67
Multiple thread cutting on an engine lathe, 298–301
Multiple turning, 310, 311

Nominal size, 40
Nose radius, 145, 146

Offset center method of taper turning, 229, 230
Optical comparators, 266–268
Overhead turning, 310

INDEX 397

Pitch, formula for thread, 248
Planer gage, 39, 40
Planing, 5
Plug and ring gages for tapers, 222, 223
 for threads, 262–264
Point angles of drills, 70
Polishing on a lathe, 186
Power operated hacksaw, 364
 feed rate for, 366
 saw blades for, 364–366
Precision, 41
Precision gage blocks, 34–38
Precision gage, universal, 39, 40
Precision hole location methods, buttoning, 337–341, 356–359
 jig boring, 344–355
 jig grinding, 356–360
 layout and drilling, 330–335
 layout, drilling, and boring, 336, 337
Precision roll method for measuring tapers, 225, 226
Production lathes, 317, 318
Protractor, bevel, 19, 20
 vernier bevel, 25–27
Pullouts, 163, 164

Radial drilling machine, 64, 65
Rake angles, 141–143
Readout system, digital, 186
Reamer holder, floating, 214, 313
Reamers, 77–80
 cutting speed, 79
 feed rate, 79
 floating blade type, 313
 hand, 79
 machine, 77–79
 rose, 79
 stock removal, 79
 taper, 79, 80
Reaming, basic principles of, 77–79
 feed rate for, 79
 on a drill press, 101, 102
 on a lathe, 214
 on a turret lathe, 313
 taper, 232–234
Relief angles, 141
 of drills, 69, 70
Rockwell hardness, 127
Roller bar turner, 314
Rose reamers, 79
Rough and finishing turning, 168
Rules, precision steel, 14–16

Saw blades, for power hacksaws, 364–366
 for vertical contour bandsaws, 371–378
 welding band, 376–378

Saw guides, adjustment of, 370, 371
Sawing, friction, 383–387
Sawing machines, 364–371
 band type cutoff, 366, 367
 power hacksaws, 364
 vertical contour band, 368–371
Saws, hole, 387, 388
Screw thread micrometer, 272, 273
Screw threads (see *Threads*)
Sensitive drilling machine, 61–64
Servo-directional feed control for contour band sawing, 382, 383
Setups for drilling, 92–99
Shaping, 5
Shoulder turning, 186–189
Side-cutting-edge angle, 143–145
Sine bar, 223, 224
 method of measuring tapers, 223, 224
Single-point tool grinding, 153–157
Slide tool, turret lathe, 311
Spade drill, 83, 84, 312
Speeds for drilling, 90–92
Spot drill, 352
Spot facers, 80
Squares, 19–21
 combination square, 19
 cylindrical square, 21
 master precision square, 20
 method of checking squareness of, 20, 21
Square threads, 258
Start drill, 311
Start drilling, 100, 101, 211, 311, 312, 336, 340, 352, 353
Steady rest, 122, 123, 172, 191, 192
Stock stop and start drill, combination, 311, 312
Surface finish on the workpiece, 152, 153, 160
Surface gage, 38, 39
Surface grinding, 12

Tap drill size, 81, 82
Taper attachment, 226–228
 method of taper turning, 226–229
Taper boring, 232–234
Taper reamers, 79, 80
Taper reaming, 232–234
Tapers, American Standard Machine, 221
 Brown & Sharpe, 220
 calculations for, 221, 222
 Jarno, 220
 measuring and gaging, 222–226
 Milling Machine, 221
 Morse, 220
 plug and ring gages for, 222, 223

INDEX

Tapers, precision roll method for measuring, 225, 226
 sine bar method of measuring, 223, 224
Taper thread cutting on an engine lathe, 297, 298
Taper thread turning, 297, 298
Taper turning, 226–232
 compound rest method, 230–232
 offset center method, 229, 230
 taper attachment method, 226–229
 taper threads, 297, 298
Tap holder, releasing type, 313
Tapping, 80, 83, 102–105, 313
 cutting speed, 102–104
Taps, 80–83
Telescoping gage, 19
Test indicators, dial, 33, 34
Thread chasing dial, 286–288
Thread cutting on an engine lathe, 275–301
 aligning tool in existing thread groove, 293
 boring internal threads, 294–297
 change gear ratios, 280–285
 compound rest setting, 285, 293, 294
 cutting American Standard threads, 289–293
 cutting Acme threads, 297
 cutting left-hand threads, 293, 294
 cutting metric threads, 283, 284
 cutting multiple threads, 293–301
 cutting speed, 285, 286
 cutting taper threads, 297, 298
 cutting worm threads, 284, 285
 setting the thread cutting tool, 289, 294
Thread cutting tools, 275–280
Thread pitch, formula for, 248
Threads, American National Thread System, 255
 American Standard Acme Threads, 256–258
 American Standard Buttress Screw Thread, 259
 American Standard Microscopic Objective Threads, 256
 American Standard Taper Pipe Threads, 258, 259
 American Standard Unified Miniature Screw Threads, 256
 American Standard Unified Thread Classes, 254, 255
 American Standard Unified Thread Designations, 255
 American Standard Unified Thread Form, 250–252
 American Standard Unified Thread Series, 253, 254

Threads, American Standard Unified Thread System, 252, 253
 British Association Standard (BA), 260
 British Standard Fine (BSF), 259, 260
 British Standard I.S.O. Metric Threads, 260, 261
 British Standard Whitworth, 259, 260
 definitions of terms, 248–250
 interference fit threads, 256
 International Metric Thread System, 261, 262
 mating part as a gage for, 264
 measuring and gaging, 262–273
 optical comparators for, 266–268
 plug and ring gages for, 262–264
 screw thread micrometer for, 272, 273
 snap gages for, 265, 266
 square threads, 258
 standard forms and systems, 250–262
 three wire method of measuring, 268–272
 toolmakers microscope for, 268
Three wire method of measuring screw threads, 268–272
Tolerances, bilateral and unilateral, 41
 transfer of, 324–328
Tool grinding, single point, 153–157
Tool holders for single-point tools, 146–148
Tooling, turret lathe, 309–317
Tool life, 128, 132, 133, 135, 162, 165
Toolmakers buttons, 337, 338
Toolmakers microscope, 268
Tool materials (see *Cutting tool materials*)
Tool steel as a cutting tool material, 137
Tracer attachment, 123–125
Trial cut procedure, 183–185
Turning, 168–198
 angle, 230–232
 basic operation of, 168
 center, 180–186
 feed rate in, 132–135, 182, 197, 198
 multiple, 310, 311
 overhead, 310
 principles of, 4, 5
 rough and finish, 168
 shoulder, 186–189
 taper, 226–232
 with part in chuck, 209, 210
Turret drilling machine, 65, 66
Turret lathe, 304–309
 boring, 311
 cutting off, 315
 drilling, 310–312
 external threading, 313

Turret lathe, multiple turning, 310, 311
 operation planning, 315, 316
 overhead turning, 310
 reaming, 313
 roller bar turner, 314
 start drilling, 311, 312
 stock stop, 311, 312
 tapping, 313
 tooling, 309–317 (see *Turret lathe tooling*)
 vertical, 319–321
Turret lathe tooling, 309–317
 combination stock stop and start drill, 311, 312
 cutter holders, 310
 floating reamer holder, 214, 313
 multiple turning head, 310
 rear cutter block, 315
 releasing type tap holder, 313
 roller bar turner, 314
 self-opening die head, 313
 slide tool, 311
 square turret, 311
Twist drill, 67–70

Unilateral tolerance, 41

Unit horsepower, 194, 195
Universal precision gage, 39, 40
Upright drilling machine, 61

V-blocks, 48, 98, 246
Vernier measuring instruments, 22–27
 bevel protractor, 25–27
 calipers, 25
 depth gage, 25
 gear tooth caliper, 279
 height gage, 25
Vernier scale reading, 22–27
 25-division vernier, 22–24
 50-division vernier, 24, 25
 vernier bevel protractor, 25–27
Vertical boring mills, 304, 318–321
Vertical contour band saw, 368–371
Vertical lathes, 304, 318–321
Vertical turret lathes, 319–321

Welding band saw blades, procedure for, 376–378
Witness marks, 46, 100
Worm thread cutting on an engine lathe, 284, 285